The Open University

Science Foundation Course Unit 5

THE STATES OF MATTER

Prepared by the Science Foundation Course Team

THE OPEN UNIVERSITY

A NOTE ABOUT AUTHORSHIP OF THIS TEXT

This text is one of a series that, together, constitutes *a component part* of the Science Foundation Course. The other components are a series of television and radio programmes, home experiments and a summer school.

The course has been produced by a team, which accepts responsibility for the course as a whole and for each of its components.

THE SCIENCE FOUNDATION COURSE TEAM

M. J. Pentz   (Chairman and General Editor)

F. Aprahamian	(Editor)	J. McCloy	(BBC)
A. Clow	(BBC)	J. E. Pearson	(Editor)
P. A. Crompton	(BBC)	S. P. R. Rose	(Biology)
G. F. Elliott	(Physics)	R. A. Ross	(Chemistry)
G. C. Fletcher	(Physics)	P. J. Smith	(Earth Sciences)
I. G. Gass	(Earth Sciences)	F. R. Stannard	(Physics)
L. J. Haynes	(Chemistry)	J. Stevenson	(BBC)
R. R. Hill	(Chemistry)	N. A. Taylor	(BBC)
R. M. Holmes	(Biology)	M. E. Varley	(Biology)
S. W. Hurry	(Biology)	A. J. Walton	(Physics)
D. A. Johnson	(Chemistry)	B. G. Whatley	(BBC)
A. B. Jolly	(BBC)	R. C. L. Wilson	(Earth Sciences)
A. R. Kaye	(Educational Technology)		

The following people acted as consultants for certain components of the course:

J. D. Beckman	R. J. Knight	J. R. Ravetz
B. S. Cox	D. J. Miller	H. Rose
G. Davies	M. W. Neil	
G. Holister	C. Newey	

The Open University Press Limited
Walton Hall, Bletchley, Bucks

First Published 1971
Copyright © 1971 The Open University

Designed by The Media Development Group of the Open University

Printed in Great Britain by
Oxley Press Ltd,
London and Edinburgh

SBN 335 02030 5

Open University courses provide a method of study for independent learners through an integrated teaching system, including textual material, radio and television programmes and short residential courses. This text is one of a series that make up the correspondence element of the Science Foundation Course.

The Open University's courses represent a new system of university level education. Much of the teaching material is still in a developmental stage. Courses and course materials are, therefore, kept continually under revision. It is intended to issue regular up-dating notes as and when the need arises, and new editions will be brought out when necessary.

Further information on Open University courses may be obtained from The Admissions Office, The Open University, P.O. Box 48, Bletchley, Buckinghamshire.

Contents

Table A

List of Scientific Terms, Concepts and Principles used in Unit 5

Taken as pre-requisites			Introduced in this Unit			
1 Assumed from general knowledge	**2** Introduced in a previous Unit	Unit No.	**3** Developed in this Unit	Page No.	**4** Developed in a later Unit	Unit No.
density	kinetic energy	3–4	gaseous state	7	mole	6
			Boyle's Law	9		
pressure	momentum	3–4	ideal gas	9		
			ideal gas law	9		
atmospheric pressure	mass	3–4	temperature	9		
			absolute zero	10		
			Kelvin (scale or degree)	10		
(above terms also in	velocity	3–4	$T=\dfrac{\bar{P}}{R}V$	10		
Appendix 1)	M K S units	1–4	triple point of water	10		
			Centigrade (scale or degree)	11		
melting	conservation of momentum		Fahrenheit (scale or degree)	11		
boiling	and energy		Zeroth Law of Thermo-dynamics	12		
	(First Law of		elastic collision	15		
metals	Thermo-		gas pressure	15		
(malleability of)	dynamics)	3–4	inelastic collision	15		
			Boltzmann constant	19		
	Newtonian		$T \propto \bar{v^2}$	19		
average	mechanics	3–4	Avogadro's Law	20		
			Maxwell distribution	20		
			cohesion	22		
	In MAFS		intermolecular force	22		
			Second Law of Thermo-dynamics	22		
	vectors		liquid state	23		
			crystal imperfections	25		
	simple algebraic operations		crystal structure	25		
			solid state	25		
	Pythagoras' theorem		sublimation	26		
			crystal grains	27		
			glasses	27		
			entropy (probability/order)	33		
			fusion (latent heat of)	33		
			vaporization (latent heat of)	33		
			$\Delta S = L/T$	34		
			Trouton's Rule	34		
			Third Law of Thermo-dynamics	35		

Any scientific terms used in this Unit but not listed are marked thus † and defined in the glossary.

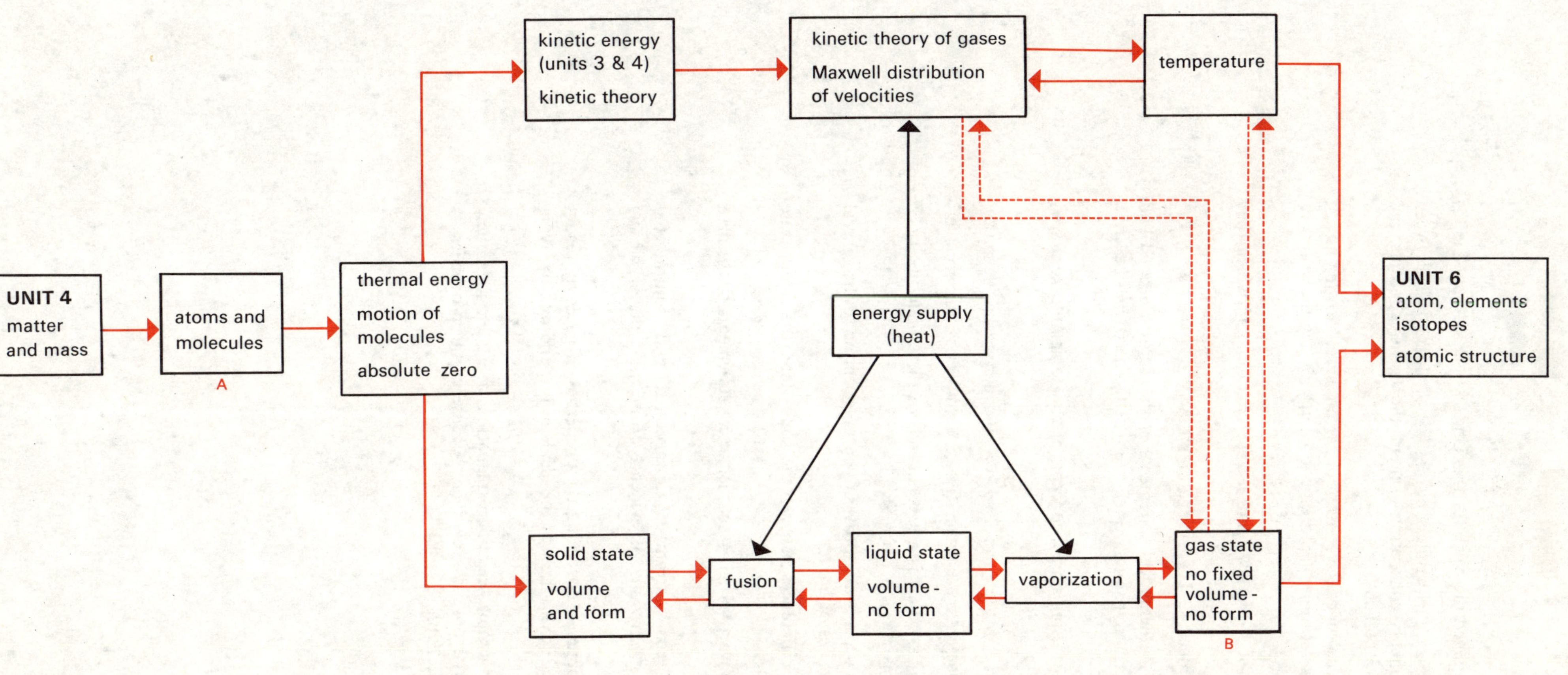

Notes:
1. To avoid confusion in the diagram, the laws of Thermodynamics are omitted. 2. The Unit starts at points A and B simultaneously

Objectives

When you have completed the work for this Unit, you should be able to:

1 Define, or recognize adequate definitions of, or distinguish between true and false statements concerning each of the terms, concepts and principles in column 3 of Table A.

2 Give a brief description of the concept of gas pressure, showing how it is derived from consideration of the kinetic energy of gas molecules or atoms.

3 State briefly the following laws, and list the conditions in which they are valid:

> Boyle's Law, ideal gas law, Avogadro's Law.

4 Give simple non-mathematical statements of the Zeroth, Second and Third Laws of Thermodynamics. Such statements need *not* be in a completely general form.

5 Derive Boyle's Law using simple concepts of Newtonian mechanics, in the form $PV = \frac{1}{3} nm\bar{v}^2$.

6 Define temperature in the form $T = \frac{P}{R}.V$ and derive this definition by considering the elementary mathematics of a simple gas thermometer.

7 Show that Avogadro's Law is a consequence of the kinetic theory of gases by using the ideal gas law $PV = nkT$.

8 List three ways in which water proves to be an exception to the general rules about the properties of liquids.

9 List the essential differences between the solid state, the liquid state and the gaseous state.

10 Interpret and convert verbal $\rightleftharpoons$ mathematical statements in the area of the Unit.

11 Distinguish between warranted, unwarranted and contradicted conclusions from a given body of evidence from the Unit or closely connected with it.

12 Solve simple problems concerning the properties of gases by recognizing the relevance of Boyle's Law or the ideal gas law and by employing them accurately.

13 Suggest or select simple scientific explanations for phenomena new to you, but which arise almost exclusively because of the properties of matter described in the Unit.

5.1 Introduction

For many centuries men have wondered whether it is possible to sub-divide matter *ad infinitum*. The Greek natural philosophers concerned themselves with this question but it was not until the time of the English chemist Dalton that the atomic hypothesis of matter was formulated in its modern dress. At this stage in the course, Dalton's evidence for his hypothesis will be omitted; it will be treated in the radio programme of Unit 6. For the first part of this present Unit the atom will be considered to be a particle of extremely small diameter, as it was by Dalton. It is now known that atoms are roughly 10^{-10} m in diameter (Units 6 and 30). These atoms were regarded by Dalton as indivisible, and for the moment his picture will be accepted. Later on you will see that they are not in fact indivisible and that twentieth-century science is vitally concerned with their internal structure (Unit 6 and also Unit 2). Dalton showed that atoms, which are the smallest particles of a chemical element, can combine in specific ways and in specific numbers to form molecules, which are the smallest particles of a chemical compound. Again you are referred forward to Unit 8 for details.

A molecule, as you will see in Unit 8, is a specific group of atoms which, due to the particular forces in operation, forms a stable entity—work must be done on the molecule to separate the atoms in it.

However, many materials are not stable in molecular form. For example, while gaseous oxygen exists as molecules of two atoms (O_2), gaseous copper consists of individual copper atoms or arbitrary groups of copper atoms.

For generality, in considering the states of matter, molecules will be discussed without further qualification until section 5.5 (the solid state). As you will see in this Unit, many properties can be derived from the simple picture; in particular the different states of matter (solid, liquid and gas) can be understood in terms of the different sorts of interaction between their constituent molecules or groups of atoms.

Do you understand what is meant by solid, liquid and gas?

Answer this question by making a list of the external properties of (i) a solid—say a lump of ice, or a block of copper; (ii) a liquid—water, or mercury; (iii) a gas—steam or air.

Further on in the Unit you will be asked to look at your list again, and amend it as necessary (see self-assessment question 10).

You will also see, in textbooks and so on, references to the 'solid state', or 'solid phase', the 'liquid state/phase' or the 'gaseous state/phase'. Do

gaseous state

not let these forms of words confuse you, they mean just what they say;
the state of being a solid, and so on.

(i) Do you know the meaning of the terms *density* and *pressure*?

Yes □   No □

(ii) Do you know what is meant by 'a pressure of one atmosphere' or 'atmospheric pressure'?

Yes □   No □

(iii) Calculate the pressure at the bottom of a lake 30 m in depth given that $g = 9.8$ m s^{-2} and the density of water is 10^3 kg m^{-3}. (Calculate in newton metre^{-2}.)

If you had any difficulty with question (iii), or if you answered 'No' to questions (i) and (ii), turn to the red-page Appendix 1 on 'Densities and Pressures'. If you had no difficulty, proceed straight to the following section of the course text.

The pressure is given by $P =$ height $\times$ density $\times$ acceleration due to gravity, so the excess pressure at the bottom of the lake is $30 \times 10^3 \times 9.8$ N m^{-2}. But the pressure at the surface of the lake is already one atmosphere, which will support a column of 0.76 m of mercury or of 10.3 m of water. So the *total* pressure at the lake bottom is given by $(30 + 10.3) \times 10^3 \times 9.8$ N m^{-2}, say 3.95×10^5 N m^{-2}.

5.2 Temperature, and the Ideal Gas Law

You are, of course, familiar with the concept of temperature. In this section you will see how temperature can be *defined* by considering the properties of a gas.

In the seventeenth century, Robert Boyle showed that the product of the pressure, P, acting on a given amount of gas, and the volume, V, occupied by that amount of gas, is a constant for many gases provided the temperature remains constant. This is Boyle's Law, expressed either as $PV =$ constant or as $P_1V_1 = P_2V_2$ (where the P_1V_1 refers to one pair of values for a given mass of gas, P_2V_2 to the second pair of values for the same mass). You will read later about conditions when Boyle's Law is *not* true (Section 5.3.5).

Boyle's Law

It was subsequently shown that the constant in Boyle's Law depends on the temperature, and indeed increases uniformly with temperature.

To measure any temperature it is necessary to have some substance with a property which changes with temperature, and to *define* temperature in terms of change in this 'thermometric property' of the 'thermometric substance'. Historically temperature was measured with mercury-in-glass thermometers, using the expansion of the metal mercury (which is liquid at normal temperatures) contained in a glass capillary (a very narrow tube). You have surely used such a thermometer to take a body temperature during sickness.

Later it became clear that mercury-in-glass thermometers, although adequate as secondary standards* of temperatures for normal laboratory work, were not satisfactory as a primary standard* for the definition of temperature, because it was impossible to be sure that thermometers made in different places would give identical readings. For this reason the properties of gases were chosen for a primary definition of the scale of temperature. Although 'gas thermometers' are more difficult to set up and use, they are much less dependent on local variations in conditions or variations in the properties of the thermometric substances.

Temperature is *defined* by the equation $PV = RT$ where P, V, T are the pressure, volume and temperature of a chosen amount of gas and R is a constant. For 1 mole** of a gas, R is a universal constant for all gases and has the value 8.3 joule per degree Kelvin (see section 5.2.1). R may thus be written 8.3 joule mole^{-1} degree^{-1} or 8.3 J mol^{-1} K^{-1}.

temperature

The law $PV = RT$ is called the 'ideal gas law' or sometimes the 'perfect gas law'. Many gases which can be observed in the laboratory do not obey this law precisely; they 'deviate' from the ideal gas law, particularly as the pressure increases, to an extent which depends on the particular gas and is very different, for example, for nitrogen and carbon dioxide. A gas which obeys the law precisely is, by definition, an 'ideal gas' (or 'perfect gas'). In general, all gases tend to behave as ideal gases as the pressure on them decreases, and it may be said that at infinitely low pressure all gases behave as ideal gases. Under normal laboratory conditions nitrogen is much more nearly an ideal gas than is carbon dioxide. Gases

ideal gas law

ideal gas

---

* *A primary standard is one by which a given quantity is defined—the standard block of platinum for the kilogramme, for example. A secondary standard is a more convenient form for laboratory use—a box of weights, for example.*

** *A mole is a unit of quantity related to a molecular weight, which you will learn about in Unit 6.*

also behave as ideal gases even at very high pressures, provided that the temperature is sufficiently high. Such conditions prevail in the centre of the sun and other stars.

It is possible to predict the deviations from ideal gas behaviour for a given gas at given temperature and pressure. You will not be concerned with the details of this in this year's course, but you will learn some of the reasons for these deviations in section 5.3.5 below.

5.2.1 Scales of temperature

A gas thermometer is arranged to work so that a given amount of gas is always held either at a constant volume or at constant pressure, and the change in temperature is then measured in terms of the change in pressure or volume respectively (see Fig. 1). From the ideal gas law, in the first case, $P=(R/\overline{V})T$ where $\overline{V}$ is the constant volume of the gas, so that P is proportional to T at constant volume. Similarly, in the second case, $V=(R/\overline{P})T$ where $\overline{P}$ is the constant pressure on the gas, so V is proportional to T at constant pressure. Notice that the form of the ideal gas law implies that as P (in the first case) or V (in the second case) approaches zero, T will also approach zero. The law thus implies an Absolute Zero of temperature, a fact that was early recognized by scientists investigating the change of volume of a gas with change of temperature (at constant pressure).

To use the ideal gas law to define temperature, it is necessary to choose some physically reproducible condition, and to assign a temperature value to that condition. The chosen point is the triple point of water. This is the one and only temperature at which solid water (ice), liquid water and gaseous water (water vapour) can exist in equilibrium with each other without, for example, the ice melting or the liquid vaporizing. The condition is technically difficult to bring about, but is chosen for its uniqueness. It occurs at a specific pressure, in fact $\sim 6 \times 10^{-3}$ atmospheres,* though you need not remember the value. By international agreement the value 273.16 K is assigned to this temperature. (See below for the meaning of the K.) This temperature is very close to the melting point of ice under normal atmospheric pressure (273.00 K at a pressure of 0.76 m of mercury). Historically the melting point is a thermometric fixed point.** You are unlikely to be concerned with very small temperature differences for present purposes, so the temperatures of these two points may be taken as identical and may be called the fixed point temperature 273 K. Then any unknown temperature can be found from the relations:

$$\frac{T}{273} = \frac{P_T}{P_{273}}$$

$$\text{or } \frac{T}{273} = \frac{V_T}{V_{273}}$$

where P_T, P_{273} are the pressures (at constant volume) measured at the unknown temperature and at the fixed point, and V_T, V_{273} are the volumes (at constant pressure) similarly measured. The unit of temperature defined in this way is called the kelvin, or formerly the degree Kelvin, hence the symbol 273 K, etc.

gas thermometer

$T=\dfrac{\overline{V}}{R}P$

$T=\dfrac{\overline{P}}{R}V$

Absolute Zero of temperature

triple point of water

Kelvin scale

* We shall use the symbol $\sim$ to signify 'approximately' or 'about'.

** Early mercury thermometers were calibrated at two 'fixed points', the freezing point and the boiling point of water under a pressure of one atmosphere. For details see elementary text books.

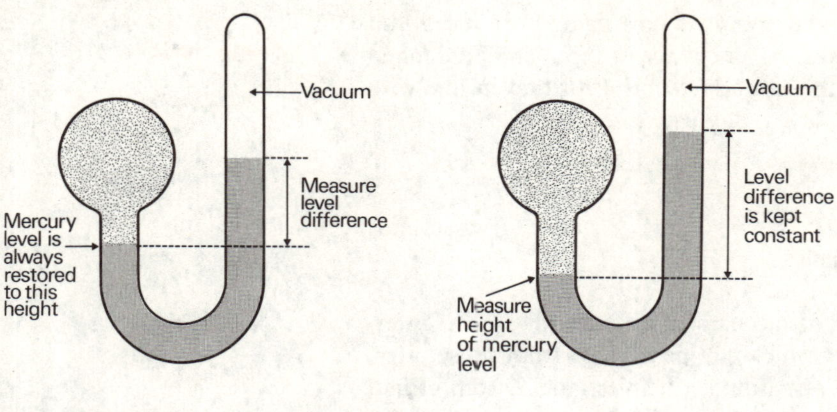

Figure 1 (a)
Constant-volume gas thermometer

(b)
Constant-pressure gas thermometer

These are purely diagrammatic, and do not represent any practical form of gas thermo-meter. In the constant-volume thermometer, the gas contained in the bulb is kept at constant volume, since the mercury is always adjusted to the same level in the left-hand tube. The difference in mercury levels gives the pressure, which will vary with temperature. In the constant-pressure thermometer, the difference in mercury levels is always adjusted to a constant value, and then the gas volume (which will vary with temperature) is given by the mercury level in the left-hand tube.

The right-hand tube is shown as containing a vacuum above the mercury. Sometimes this tube is open to the atmosphere, but then due allowance must be made for atmospheric pressure, and for any variation in atmospheric pressure, during an experiment.

In the development of science, other temperature scales have arisen and two of these are still in use today. In the Centigrade (or Celsius) scale the *size* of the degree is identical to that in the Kelvin scale. The melting point is called 0° C, so that absolute zero becomes −273° C. The boiling point of water (under a pressure of one atmosphere) is 100° C (this point was used as a fixed point in defining the scale and is 373 K on the Kelvin scale). In the Fahrenheit scale, the size of the degree is only 5/9 as large as the Kelvin. Absolute zero is −459° F, the melting point 32° F and the boiling point 212° F. Scientists all over the world express temperatures in kelvin (see Fig. 2).

Centigrade scale

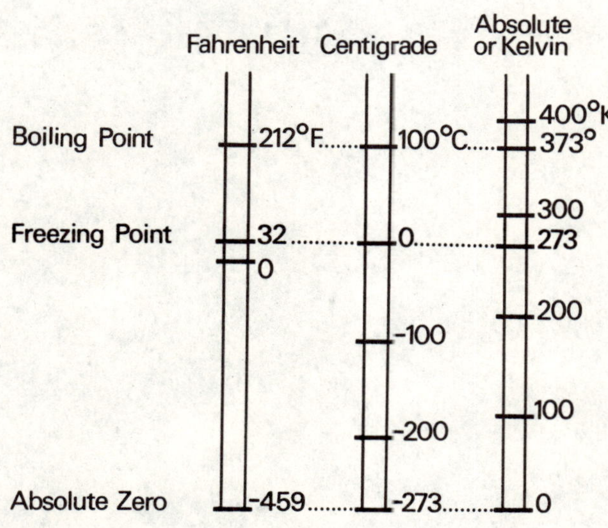

Figure 2   The Fahrenheit, Centigrade and Kelvin temperature scales side by side, showing their equivalent points.

11

The shift from Kelvin to Centigrade, the scale now being introduced into Britain for day-to-day use, involves only a change of zero. The Fahrenheit scale, now being abandoned in this country to bring Britain in line with Continental practice, persists in day-to-day use in the U.S.A.

5.2.2   The Zeroth Law of Thermodynamics

Thermodynamics is the analysis of mechanical and thermal (heat) interactions, and of the consequences which can be derived from those interactions. The subject is founded on four principles, the Zeroth, First, Second and Third Laws of Thermodynamics. You have already met one of these principles in Units 3 and 4 (the principle of the conservation of energy, which is also the First Law of Thermodynamics). In the present Unit you will meet the other three principles.

The curious system of numbering the laws of thermodynamics arose historically. The First and Second Laws had been formulated, and named, before it was realized that for logical completeness a prior condition, or Law, was needed. The Zeroth Law was not written by Dr Zero. . . .

The Zeroth Law is about thermal equilibrium. Two objects are in thermal equilibrium when no heat energy flows from one to the other if they are placed in contact. This is another way of saying that they are at the same temperature. If a third object is in thermal equilibrium with the first object, then the Zeroth Law states that it must automatically be in thermal equilibrium with the second (thus all three objects are at the same temperature).

To illustrate this principle, consider a carton of milk taken from the doorstep and an egg taken from a shopping bag, both placed on the same shelf in your refrigerator, but not in contact. Each will lose heat energy to the refrigerator, each will eventually come to the thermal equilibrium with the refrigerator. If now the egg is cracked and put in the carton of milk, still in the refrigerator, no heat will flow from egg to milk or from milk to egg, because there will be no temperature difference between them to cause heat energy to flow.

This principle is self-evident and needs no further justification.

Question 1 (*Objective 12*)

A gas cylinder, at 17° C, is filled with gas at a pressure of 250 atmospheres.

What will be the pressure if the temperature increases to 37° C? (Select from a—e):

        (a) 273 atmospheres

        (b) 256 atmospheres

        (c) 268 atmospheres

        (d) 245 atmospheres

        (e) 261 atmospheres

Question 2 (*Objective 1*)

Calculate the Fahrenheit equivalents of (i) 17° C and (ii) 37° C. (Hint— remember between melting point and boiling point 100 Centigrade degrees are equivalent to 180 Fahrenheit degrees. Remember also that melting point is 0° C and 32° F.) Select from a—e for (i) and from f—j for (ii).

(i) (a) 44° F	(ii) (f) 67° F
(b) 75° F	(g) 99° F
(c) 63° F	(h) 48° F
(d) 51° F	(i) 105° F
(e) 86° F	(j) 112° F

Question 3 (*Objective 12*)

Bubbles of gas of average diameter 1 cm escape from a wreck 72 m under the surface of the sea. Assuming that the temperature of the water is constant all the way down to the wreck, what will be the (approximate) average diameter of the bubbles when they reach the surface? (Select from a—e.)

(a) 1 cm	(d) 5 cm
(b) 3 cm	(e) 2 cm
(c) 1.5 cm	

Question 4 (*Objective 12*)

You are pumping your car tyre with an old-fashioned 'long cylinder' type of pump. You have no pressure gauge, but observe that you can hear air passing in through the valve when the piston of the pump is one third of the way down the barrel. What, *very roughly*, is the pressure in your tyre in pounds per square inch (p.s.i.). (1 atmosphere $\approx$ 14 p.s.i.) (Select from a—e).

    (a) 14 p.s.i.   (b) 18 p.s.i.   (c) 28 p.s.i.

    (d) 21 p.s.i.   (e) 42 p.s.i.

Question 5 (*Objective 12*)

A capillary tube of uniform diameter has one end sealed off. The tube is filled with dry nitrogen gas, and the other end is closed with a small pellet of mercury. In a mixture of melting ice and water at one atmosphere pressure the length of the nitrogen column is 12.4 cm. What will be the length at the point where the mercury freezes ($-39°$ C)? (Note that this arrangement is a simple *constant pressure* gas thermometer.) (Select from a—e.)

    (a)   9.1 cm   (b) 12.2 cm   (c) 7.8 cm

    (d) 11.3 cm   (e) 10.6 cm

5.3   The Interactions of Molecules

5.3.1   Interactions with the walls primarily—the gaseous state

Before you can understand the interactions between the molecules in a gas, you must understand the difference between an elastic collision and an inelastic collision. When bodies such as billiard balls collide, they are usually heated slightly by the collision; their temperature increases. Although the principles of the conservation of momentum and the conservation of energy (Units 3 and 4) must apply to such collisions, it is clear that, because some of the energy has been turned into heat energy, the total kinetic energy of the billiard balls after the collision will not be as great as it was before. Such a collision is called an *inelastic* collision.

inelastic collision

A collision may be postulated in which no kinetic energy is lost, so that the total kinetic energy is the same before and after the collision. Such a collision is called an *elastic* collision. (Notice that 'elastic', as used here, is a technical term and does not have the usual meaning.) This is an idealized picture, but some situations can be set up where this idealization is approached in practice.

elastic collision

In the kinetic theory of gases, the modern view of the gaseous state, the molecules are assumed to be making continual elastic collisions with other molecules and with the walls of the container that holds the gas. In the collisions between molecules, energy and momentum (Units 3 and 4) are constantly exchanged between the molecules, but the total energy is conserved. In the collisions with the walls of the container, momentum will be exchanged at the walls of the container, and the rate of change of momentum represents a force on the walls (Units 3 and 4). It is convenient to talk of the force per unit area, the pressure. Thus it is said that a pressure is exerted on the walls of the container. In calculating this pressure in terms of the velocity of the gas molecules, you will have a good example of a statistical process.

gas pressure

(A statistical process is one where the bulk properties, the pressure for example, are governed by the average behaviour of a very large number of molecules, rather than the individual behaviour of a particular molecule.)

Suppose that there are some gas molecules, each molecule with mass m, inside a cubic container. Now there will be a constant exchange of energy between the molecules owing to collisions between them.

Do you expect that this constant energy exchange will act so that molecules of high energy will gain yet more energy, and molecules of low energy will lose some of the energy which they have?

No, it is surely your experience that in any sort of collision between bodies of roughly equal weight (motor accidents, collisions of running children, billiard balls, etc.) the faster body tends to slow down (or lose kinetic energy) and the slower to speed up (or gain kinetic energy). See, for example, Figure 3 taken under stroboscopic flash (Unit 3) which makes the effect clear.

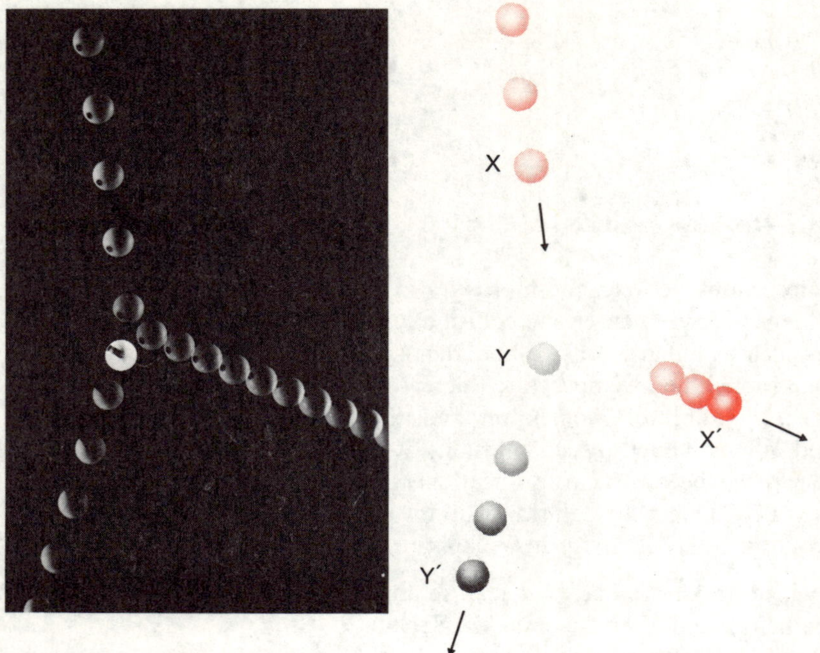

Figure 3 The incoming ball X strikes the stationary ball Y. Photographs are taken at constant time intervals, so that, in the left-hand figure, the velocities of the balls can be inferred from the distances between successive images of the balls. Clearly ball X after the collision is slowed down (X′) and Y is speeded up (Y′) (from zero velocity).

So the continuous collision of molecules can be expected to result in all the molecules having the same average energy. If the cube is very large, a molecule at the top of the cube will have a gravitational energy which is appreciably different from the (gravitational) energy of a molecule at the bottom. This is called the gravitational potential energy (Units 3 and 4). In the present derivation it will be assumed that such changes in gravitational potential energy are negligible in comparison with the kinetic energy of the molecules. The condition of the same average energy means the same average *kinetic* energy. Since all the molecules have the same mass and, on the average, the same kinetic energy, they will have some average velocity, which will be called $\bar{v}$.

Do you expect the process of energy transfer to produce finally a situation where all the molecules have exactly the same energy?

No. A distribution of energies will be set up. Some molecules will, momentarily, have more than the average energy, some less. At a given temperature, the total number of molecules which have a certain energy (for example, twice the average) will be constant, but a given molecule may at one moment have less than the average energy and at the next moment more. So the total distribution will be the same at a given temperature, but the individual molecules will be changing their positions in the distribution.

A further interesting point arises here, partly of physics but partly perhaps of philosophy. Why should the effect of constant collisions between the molecules be to give all the molecules the same average energy and to set up a fixed distribution of energies? Why should it not rather make the fast molecules faster and the slow molecules slower, for example? These points will be discussed below (see Maxwell's distribution of velocities, and the Maxwell Demon, Section 5.3.4).

First, however, *assume* that the molecules have an average kinetic energy (and thus an average velocity, $\bar{v}$) and follow the argument which relates this velocity to the properties of the ideal gas.

5.3.2 The kinetic theory and the ideal gas law

Suppose that the gas which you are considering consists of n molecules
each of mass m. Suppose that the gas is contained in a cube with a side
a metres long and define x, y, and z axes along the sides of the cubic
container. Then any molecule travelling in any direction can be considered
in terms of its velocities along those three axes v_x, v_y, and v_z, where the
vector sum (see *MAFS*, Section 4.D and also Units 3 and 4) of v_x, v_y, and
v_z is the actual velocity of the molecule, v, in any random direction.
v_x, v_y, and v_z are called the *components of velocity* in the x, y, and z
directions (see Fig. 4). Similarly the momentum of each molecule can be
considered in terms of the effects of the *components of the momentum*
along the x, y, and z axes. These momentum components are mv_x, mv_y,
and mv_z.

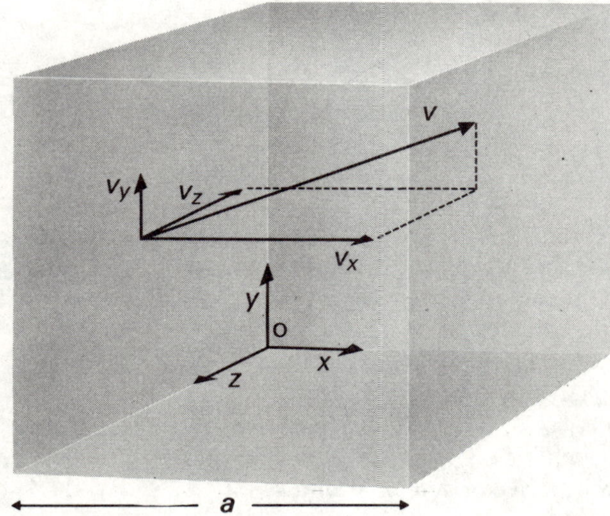

Figure 4 (a)   Velocity components of one of n *gas molecules in a cube of side* a.

Now consider the momentum exchange (and thus the force, and pressure)
during a collision between one molecule and the wall of the box, the wall
including the y and z axes, so that the momentum components mv_y and
mv_z have no effect in the momentum exchange at this face. The momentum
component of the molecule will be mv_x before the collision, and since the
collision is an elastic collision (a basic assumption of the kinetic theory)
the molecule will rebound with the same velocity reversed. The total
momentum given to the wall in this collision will be:

$$mv_x - m(-v_x) \text{ or } 2mv_x \text{ (see Fig. 4b).}$$

(The velocity after the collision is the velocity before reversed, hence $(-v_x)$).

If this momentum is given to the wall during a small time interval t (secs)
the impulsive force on the wall due to this collision is given, from Newton's
Second Law, by

$$F_x = 2mv_x/t$$

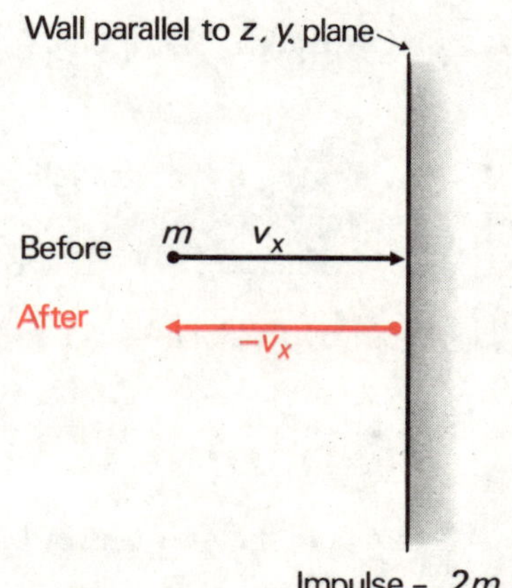

*Figure 4 (b)   A molecule colliding with
a wall parallel to the* z, y *plane transfers
a momentum* 2mv_x *to it.*

What is the physical significance of the time interval t in this equation?

In fact the time during which the momentum is exchanged is the 'time of
impact' of the particle and the wall, that is the time during which they are
in contact. However, the object of the present exercise is to get an *average*
value for the force due to all the molecules. For this purpose you might
surmise that you could use a time t which is the time between successive
collisions of the same molecule with the same wall. This is the time which
it takes the molecule to travel along one side of the box and back with
velocity v_x, or $t = 2a/v_x$. A more complete (and mathematically more

complicated) analysis would justify your surmise. The average value of the force on this wall due to this one molecule making impacts on the y–z face is thus found by eliminating t, giving:

$$F_x = \frac{2mv_x}{2a/v_x}$$

$$= mv_x{}^2/a$$

The average value of the force on this wall due to *all* the n molecules is given by:

$$F_x = nm\overline{v_x{}^2}/a$$

where $\overline{v_x{}^2}$ is the average value of $v_x{}^2$ for all the n molecules.

The average pressure, P, on this wall is, by definition, the force per unit area. Since the area is a^2, the pressure is given by:

$$P = F_x/a^2 = nm\overline{v_x{}^2}/a^3$$

$$\text{or} \qquad Pa^3 = nm\overline{v_x{}^2}$$

But a^3 is the volume, V, of the gas,

$$\text{so} \qquad PV = nm\overline{v_x{}^2}.$$

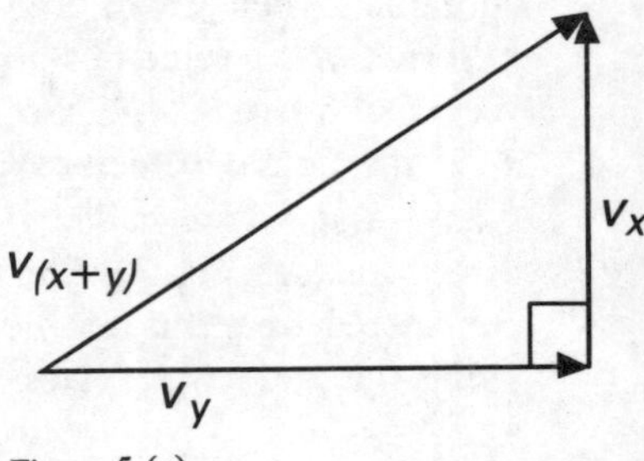

Figure 5 (a)

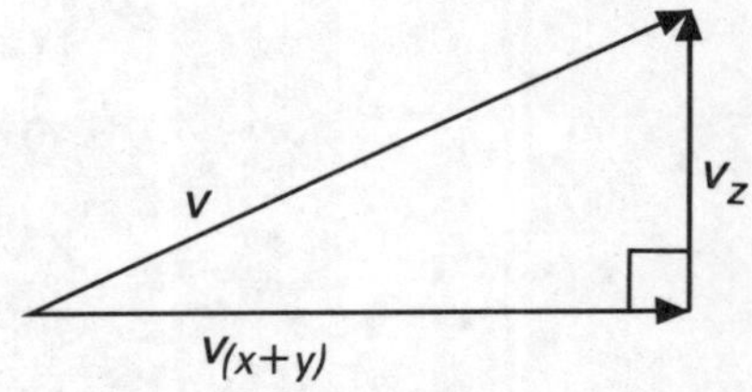

Figure 5 (b) Illustrating the vector addition of the velocity components v_x, v_y *and* v_z *to obtain the resultant velocity* v.

This is beginning to look like Boyle's Law ($PV = $ constant). The final step is to relate $\overline{v_x{}^2}$, the average value of the square of the velocity in the x direction, to $\overline{v^2}$, the average value of the squares of the individual velocities of all the molecules (this is called the mean squared velocity). To do this you can use the three-dimensional extension of Pythagoras' Theorem* applied to the three components v_x, v_y, v_z, of the total velocity v of the molecule first considered. For simplicity start with a two-dimensional picture, say with v_x and v_y. Now the vector sum of v_x and v_y can be represented in magnitude and direction by $v_{(x+y)}$ in the triangle shown in Figure 5a.

By Pythagoras' Theorem:

$$v_{(x+y)}^2 = v_x^2 + v_y^2.$$

Now draw another triangle and add $v_{(x+y)}$ to v_z (remembering that v_z is at right-angles to both v_x and v_y, and so must also be at right-angles to $v_{(x+y)}$) (Fig. 5b).

Again by Pythagoras' Theorem:

$$v^2 = v_{(x+y)}^2 + v_z^2$$

$$= v_x^2 + v_y^2 + v_z^2.$$

Now take the averages of both sides giving:

$$\overline{v^2} = \overline{v_x^2} + \overline{v_y^2} + \overline{v_z^2}.$$

(In words, the average value of the square of the velocity of the molecule is equal to the sum of the average values of the squares of the components of the velocity in the three directions x, y, z.)

But because all the molecules are moving at random in the cubic container, there is no reason why their average velocity in any one direction should be different from their average velocity in any other. So

$$\overline{v_x^2} = \overline{v_y^2} = \overline{v_z^2},$$

and since

$$\overline{v_x^2} + \overline{v_y^2} + \overline{v_z^2} = \overline{v^2}$$

each of the average squared velocities ($\overline{v_x^2}$, etc.) must be $\tfrac{1}{3}\overline{v^2}$.

** See MAFS, section 4.C.2 if necessary.*

Substituting, the final relation is obtained

$$PV = \tfrac{1}{3}nm\overline{v^2} \quad \dots\dots\dots\dots\dots\dots(1)$$

Compare this with the ideal gas law,

$$PV = RT \quad \dots\dots\dots\dots\dots\dots\dots(2)$$

You will see that the left-hand sides are equal. What does this tell you about the right-hand sides?

They must also be equal.

What, in particular, do you notice about the right-hand side of (1)?

The most striking thing is surely the form $m\overline{v^2}$. Remember, from Units 3 and 4 the expression for the kinetic energy of a particle mass m, with velocity v was $\tfrac{1}{2}mv^2$, so that $\tfrac{1}{3}nm\overline{v^2}$ is $\tfrac{2}{3}n \times (\tfrac{1}{2}m\overline{v^2})$, which is $\tfrac{2}{3}n$ of the average kinetic energy of a molecule of the gas due to its movement through the body of the gas. This can be written $\tfrac{2}{3}n\,(\overline{KE})$. Notice that movement through the body of a gas (translational motion) is not the only sort of motion possible for a gas molecule. It may also be spinning on some axis, as the Earth spins on the polar axis, or it may be vibrating in some way internally. Many molecules are dumb-bell shaped rather than spherical and may vibrate, for example along the line joining the centres, so that the ends of the dumb-bell are from moment to moment closer together or further apart. These motions will all have associated kinetic energy, but these energies are excluded in the relation given above, where the symbol $(\overline{KE})$ represents *only* the kinetic energy due to the movement of the molecule.

Notice also that in equation 1, $\overline{v^2}$ is the average of the *squares* of the velocities of the individual molecules. This is not, in fact, the same as the square of the average velocity, but it is mathematically related to that quantity. For present purposes the significant quantity is $\overline{v^2}$, because the interest is in the average kinetic energy of the molecules. So equation 1 becomes:

$$PV = \tfrac{2}{3}n\,(\overline{KE}) \quad \dots\dots\dots\dots\dots(3)$$

Comparing equation 3 with 2, you sense that there is a relationship between $(\overline{KE})$, the average kinetic energy of any one molecule, and the temperature T. The concept which is called 'temperature' is a manifestation of the average kinetic energy of the molecules, since when the average kinetic energy increases this is sensed as a 'hotter' gas. There must then be a relationship between $(\overline{KE})$ and the temperature. This relationship is one of proportionality, and is given by:

temperature as a measure of kinetic energy

$$(\overline{KE}) = \tfrac{1}{2}m\overline{v^2} = \tfrac{3}{2}kT \quad \dots\dots\dots\dots\dots(4)$$

k is called Boltzmann's constant, and in SI units is 1.38×10^{-23} J K^{-1}. There is a relationship between R and k which should be apparent after you have read section 5.3.3.

Boltzmann constant

The introduction of the constant of proportionality in the form $\tfrac{3}{2}k$ depends on a branch of classical physics called statistical mechanics. The details are beyond the scope of this Unit, but some of you will return to the point in the second-year course 'Gases, Liquids and Solids'.

Equation 4 can be regarded as an alternative mechanical definition of the Kelvin Temperature K. Indeed, some textbook writers regard this as more fundamental than the experimental definition given in section 5.2.1. Notice that k is a constant for all gases. If the molecules in one gas are heavier than those in another gas at the same temperature, $\overline{v^2}$ for the

19

heavier gas will be smaller, but $(\overline{KE})$ will still be $\frac{3}{2}kT$. Combining 3 and 4 gives:

$$PV = nkT \quad \dots\dots\dots\dots\dots(5)$$

5.3.3 Avogadro's Law* and Avogadro's number

The chemist Avogadro, studying the reactions between gases, made the hypothesis that 'equal volumes of all gases, under the same conditions of temperature and pressure, contain the same number of molecules'.

Does Avogadro's Law follow from equation 5?

Yes, because if P, V and T are equal between two gases, and k is a constant by definition for all gases, n must also be equal for the two gases.

Do you remember the value of the gas constant R for 1 mole?

$R = 8.3$ J mol^{-1} K^{-1}.

Compare equation 5,

$$PV = nkT$$

with the ideal gas law,

$$PV = RT.$$

Obviously R and nK are equivalent. For one mole, n is the special number of molecules in one mole, and this is the same for all substances. This number is called Avogadro's number N_A.

Avogadro's number

$$N_A = \frac{R}{k} = \frac{8.31}{1.38 \times 10^{-23}}$$

$$= 6.02 \times 10^{23} \quad \text{molecules per mole.}$$

5.3.4 Maxwell's distribution of velocities, the Maxwell Demon and the Second Law of Thermodynamics

Using a more general approach and more complicated mathematics, it can be demonstrated that the actual velocities of the gas molecules are

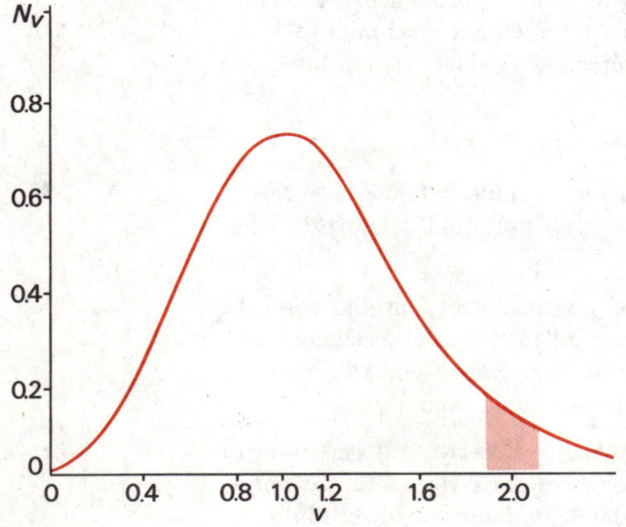

Figure 6   The Maxwell distribution of velocities, showing the number of molecules (N_v) having a particular velocity (v). The average velocity is shown as 1.0 on the v scale. The hatched area shows the number of molecules having between 1.9 and 2.1 times the average velocity. This area is about three per cent of the total area under the curve, which represents the total number of molecules of all velocities.

* *Also called Avogadro's Hypothesis.*

distributed about the average velocity according to the distribution which is called the Maxwell distribution.

The distribution is shown graphically in Figure 6. Notice that some of the molecules have velocities appreciably higher (and lower) than the average velocity. Three per cent, for example, have velocities between 1.9 and 2.1 times the average velocity. (These velocities are represented by the area hatched in on Figure 6; the total number of molecules is represented by the total area under the curve.) Figure 7 shows the effect of temperature change on the Maxwell distribution.

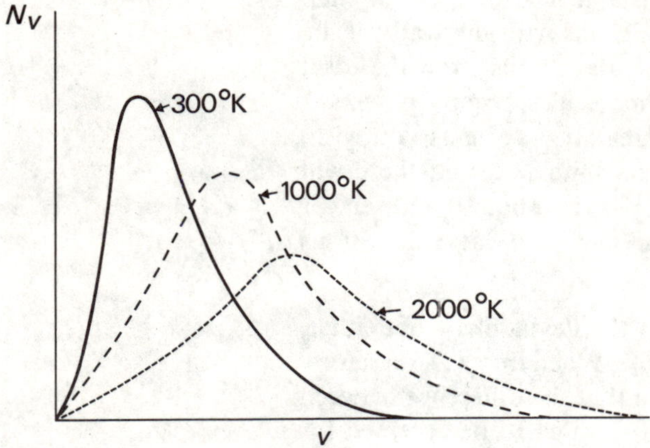

Figure 7 To show how the Maxwell distribution changes with temperature. Notice that at higher temperatures the distribution becomes broader.

Now what would happen if all three per cent of the molecules, having a velocity of about twice the average, collected at one point in the room in which you are sitting, say in the few per cent of the room volume in the neighbourhood of your head?

The temperature in the region of your head would suddenly increase (since temperature is proportional to $\bar{v^2}$) to four times the average value in the room, say from 300 K (above Absolute Zero, remember) to 1 200 K. This would certainly, and literally, boil your head. But you know by experimental observation (and by human experience over the centuries!) that this does not happen.

James Clerk Maxwell, who worked all this out, amused himself by postulating a superhuman being, the 'Maxwell Demon', who could sit by a hole in a wall that divided a gas container into two parts, A and B, and, using a sort of microscopic dustbin lid, would allow only those molecules moving at above-average velocity to pass from side A to side B, and only those moving with below-average velocity to pass from side B to side A.

'Maxwell Demon'

What would happen to the gas on the side A of the wall where the fast molecules were allowed to congregate?

The gas on side A of the wall would heat up compared with the gas on side B, because the average kinetic energy of the molecules on side A would increase, and the average kinetic energy of the molecules on side B would decrease.

Again, observation over the centuries shows that this sort of statistical improbability does not happen spontaneously. Scientists have generalized this experience in the Second Law of Thermodynamics.

One rather general statement of the Second Law of Thermodynamics is,
'Highly improbable states of affairs are not achieved except by expending
energy'. (In the case of the Demon, the energy which he expends wielding
a dustbin lid.)

5.3.5 Ideal gases and real gases

In the treatment of gases, and in the development of the ideal gas law,
$PV = RT$, the intermolecular collisions have been assumed to be elastic.
Clearly this may not be true; there are probably forces of attraction
between the molecules so that there will be a tendency for them to stick
together ('cohere') when they collide. The collisions with the wall will be
hindered by this stickiness and this will cause deviations from the ideal
gas law. It has also been assumed that the molecules occupy no space,
that they can be compressed together to as small a volume as you wish.
In fact, though, the molecules will have some finite size, and there will
come a point when the molecules are touching (as in a liquid) and further
compression will be difficult or impossible. This again will cause deviations
from the ideal gas law.

Scientists think of a real gas as one in which the deviations from either,
or both, of these effects may become appreciable. When they say, therefore,
that a gas 'behaves as an ideal gas', they mean that the difference between
actual behaviour and ideal behaviour is small enough to be accepted for
the purposes in hand. In general, all gases behave as ideal gases as the
pressure becomes small, so that the distances between gas molecules are
large. Under these conditions the intermolecular forces, which fall off
with distance, are reduced, and the volume of the molecules becomes small
compared to the total volume of the gas. Conversely at high pressures no
gases behave as ideal gases* and other equations must be introduced to
describe a real gas. These equations will not, however, concern you in
this year's course.

* *Except at very high temperatures, in stars for example, as mentioned in section 5.2.*

5.4 The Interaction of Cohesive Molecules—the Liquid State

In considering an ideal gas, it was supposed that the thermal (kinetic) energy of the molecules was sufficient to overcome any tendency which two colliding molecules might have to cohere.* In the liquid state, the molecules have sufficient thermal energy to keep them in motion but not sufficient to separate the molecules one from another. Thus the molecules cohere, but move past each other and round each other in a liquid manner. This can be compared to a barrel full of peas which are in constant motion round and past each other, but where one pea is at any moment in contact with a number of near neighbours. There is no regularity of structure; a pea that is at one instant in one corner of the barrel may at a later instant be in quite a different point in the barrel. The peas can be put into a different shaped container in which they can still occupy the same volume but have a different overall shape.

The molecules of a liquid, then, are in contact and in constant motion within a prescribed volume which is defined by the container holding the liquid (though, unlike a gas, the liquid does not always fill the container). The total volume of the liquid remains constant as long as the kinetic energy of the molecules remains the same; a liquid is a fluid** which has a definite volume but not a definite shape. This last statement could be taken as the definition of a liquid.

liquid

Just as in the case of a gas, as the temperature of the molecules increases so does their kinetic energy. In general, as the kinetic energy increases the molecules tend to stand further apart from each other while maintaining their coherence, so the average distance apart of nearest neighbours will increase with a rising temperature. Eventually, however, the kinetic energy of the molecules will be such that it is sufficient to overcome the forces of cohesion between the molecules.

What do you expect to happen then?

The liquid will boil and become a vapour or a gas.

In the gaseous state the molecules will move far apart from each other to fill any available space and there will no longer be any appreciable cohesion between molecules on impact (but see section 5.3.5 on ideal gases and real gases).

What do you expect to happen to the inter-molecular distance as a liquid cools?

The molecules will be at closer and closer average distance as the liquid cools; thus the fluidity will decrease (the liquid will not flow so readily), the molecular mobility within the liquid will decrease and the total volume of the liquid will decrease. Certainly this is the case with most liquids,

---

** The tendency of molecules to cohere can be described by the 'binding energy' of the molecules. It can be said that in the gaseous state the kinetic energy is much greater than the binding energy, while in the liquid state these two energies are about equal. See this Unit's TV programme.*

*** The word fluid means simply something which will flow, and this is your common experience of a liquid. We have deliberately avoided any quantitative definition of the fluidity of a liquid, the ease with which it will flow. See also section 5.5.*

but there is one exception to this general behaviour which is of overwhelming importance to human beings and indeed is of importance in the whole of biology. This exception is water, where between 4° C and 0° C the volume *increases* as the water is cooled. Above 4° C the volume increases as the water is heated. Thus there is a temperature of minimum volume for water, at 4° C. The case of water will be considered in some detail, after an account of the solid state.

5.5 Molecules in their Places—the Solid State

As the liquid is cooled further, it loses its fluidity and becomes a solid (it freezes). Outwardly, a solid differs from a liquid in that it is hard; it takes up and maintains a particular external shape that is determined by the constraints acting upon the liquid when it freezes (think of ice-cubes). This state of matter arises because of the increasing importance of the cohesive forces between molecules compared with their kinetic energy.* Although the molecules in a solid will still have kinetic energy of vibration, this will now be insufficient to keep them moving past and over each other, and consequently they begin to take up definite positions with respect to one another within the bulk of the material.

Are the individual units *always* molecules?

No, it was pointed out in Section 5.1 that the units *may* be definite molecules, but they may be individual atoms or arbitrary groups of atoms. For this reason two types of solid are possible: aggregates of molecules in which each molecule retains its identity, e.g. ice, and polymers (plastics); and aggregates of atoms in which the cohesive forces bond an unlimited number of atoms together, e.g. diamond (a form of carbon) and metallic elements. You can think of the latter type as very large molecules.

Now you will see in Units 9 and 10 that the cohesive forces ('bonds') between atoms or molecules can have a variety of strengths and characters and the type of 'bonding' is an important factor in determining the way in which the atoms or molecules in a solid are arranged. For the moment, think of these bonds as attractive forces between the molecules, with particular directions of maximum force.

Since the atoms or molecules attract one another they will tend to pack closely together. This is usually achieved by an arrangement in stable and regular three-dimensional arrays. Such an array is called a *crystal* and its geometric picture is a *crystal structure*. Figure 8 gives a few examples of crystal structures (p. 26). A simple example of a different sort of array is a squad of soldiers on parade for inspection.

Notice that these pictures are essentially momentary snapshots. The atoms or molecules are not stationary at their particular sites: they are constantly vibrating, and the higher the temperature the higher will be the frequency and the amplitude of vibration. If you go on heating the crystal, increasing the energy of these vibrations, more and more atoms have sufficient energy to overcome the cohesive forces holding them in regular sites and to jump into 'irregular' positions, thus leaving behind a vacant atom site. These 'irregularly placed' atoms and 'vacancies' disturb the atomic arrangement in their vicinity and are a type of 'crystal imperfection'. Several types of crystal imperfection may exist in a real crystal; you will meet some of them in this Unit, and will consider them in detail if you study the structure of matter at second and subsequent course levels. With further heating the amount of disorder in the crystal increases until a point will be reached at which the atoms or molecules begin to move freely past and round each other and the material becomes a liquid (it

---

* *Another way of saying this is that the binding energy of the molecules in a solid is much greater than their kinetic (thermal) energy.*

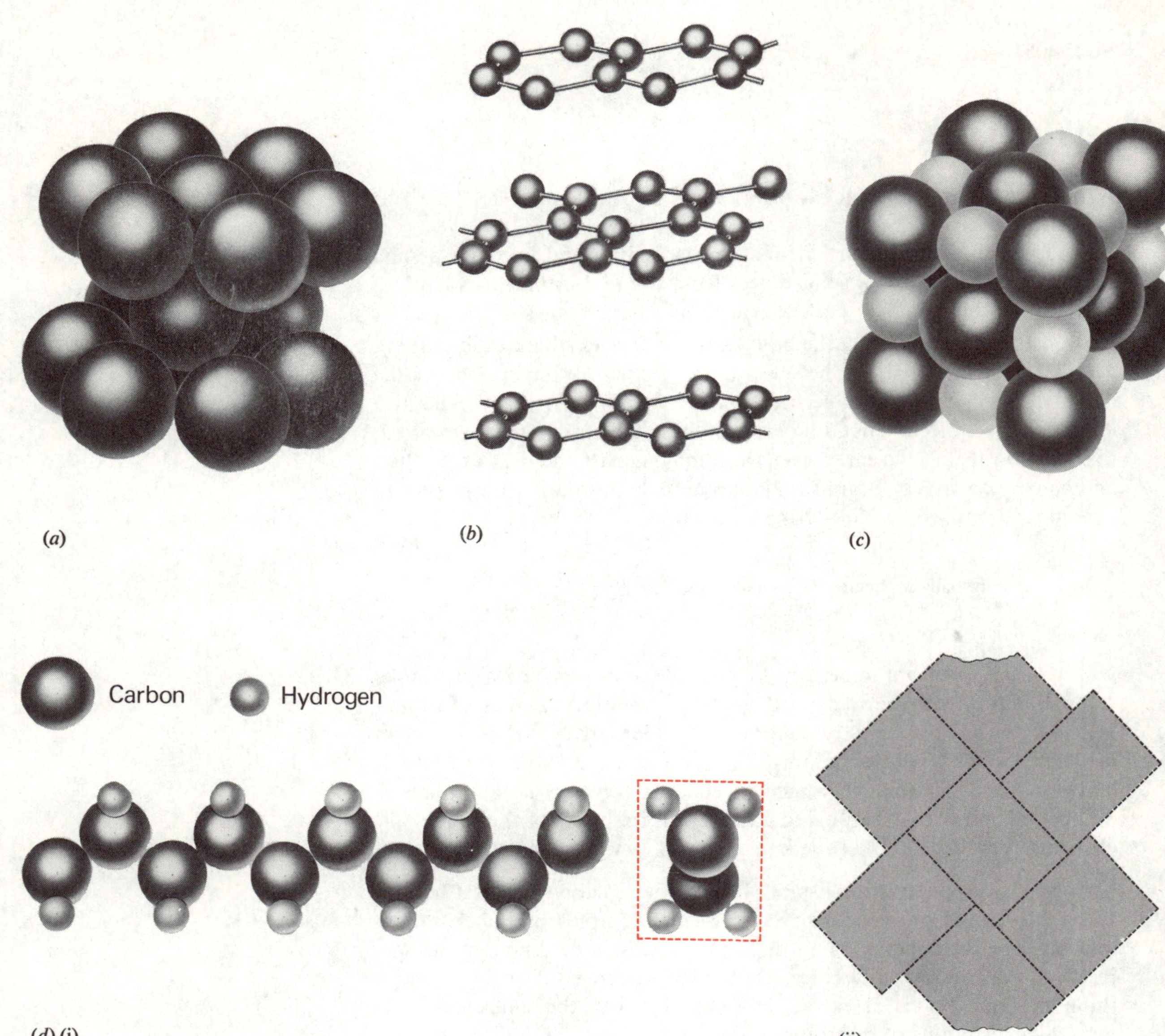

Figure 8   The illustrations are idealized, with atoms pictured as spheres. (a) shows the hexagonal structure possessed by crystals of metals such as zinc and titanium; (b) is also a hexagonal structure, of graphite (the rods represent the bonds between the carbon atoms); notice that the graphite structure consists of layers of carbon atoms. Alternate layers are displaced laterally with respect to each other; (c) is the cubic crystal structure of table salt (sodium chloride); the large spheres represent chlorine atoms, the small spheres sodium atoms; (d) shows both the molecular and crystal structures of polyethylene ('Polythene')—in (i) a side and an end-on view of the molecule, and in (ii) an end-on view showing the way the molecules pack into the crystal structure. In neighbouring chains the 'backbone' direction (the direction of the links between the carbon atoms) is different by 90°. The rectangles in (ii) are the same as the rectangle enclosing the end-on view in (i).

melts). Under some conditions, though, it is possible to go straight from the solid to the gaseous state, for example in iodine and in smelling salts (ammonium chloride) and also in the direct 'sublimation' of snow from solid to water vapour.

sublimation

Crystals are usually formed from liquids which maintain their molecular mobility down to the temperature at which they freeze; in these cases the particles can move around until they find a suitable 'regular' site on the surface of the growing crystal.

Following this, can you see that another type of solid is possible?

26

A glassy solid, where the atoms are not in regular positions, can exist.
Read on for the details of this process.

If a liquid is fairly viscous before it freezes, the atoms or molecules may
not be able to reach the regular positions characteristic of a crystal:
the material then solidifies as a mass in which the molecular arrangement
is irregular. This is called a *glass*. A major cause of viscous behaviour is
the association of neighbouring molecules to form large groups in the
liquid which can only move sluggishly. Figure 9 illustrates the structural
difference between a crystal (a) and a glass (b).

Viscosity is a measure of the resistance
to flow of a liquid (lower viscosity,
easier flow). Water is less viscous than
motor oil, which is less viscous than
treacle.

glass

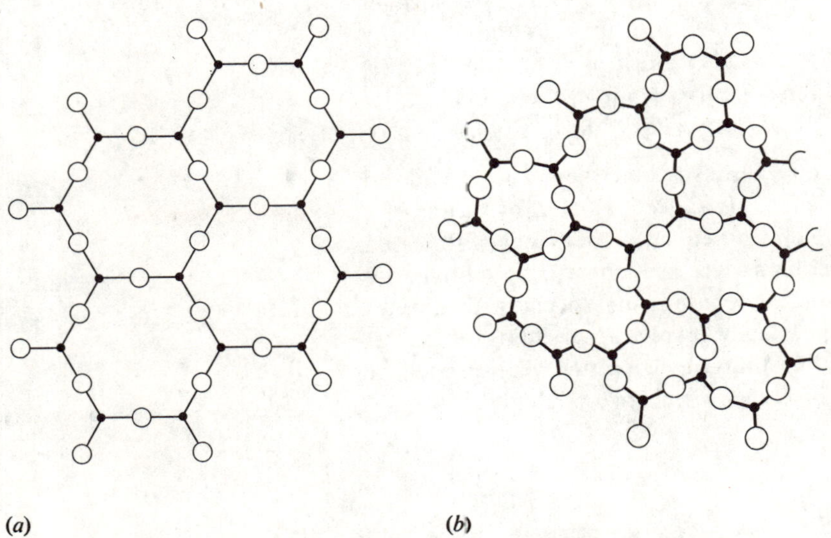

(*a*)   (*b*)

*Figure 9 (a)   A crystal, where the atoms are arranged regularly in hexagonal rings.
Each ring contains 6 atoms of the type shown as an open circle.*

*(b)   A glass, where there are no regular rings containing a standard number of atoms.
Some rings have 5 open-circle atoms, some 7, and the rings have no regular shape.*

You see that there is no long-range regularity in a glass. The glasses based
on sand are important examples of this type of solid, and many plastics
and ceramics consist of glassy and crystalline parts. Sugar is crystalline,
but can be transformed into a glass, toffee, by melting and cooling fairly
rapidly. Treacle is the viscous liquid from which the glass can be formed
and 'fudge' is devitrified toffee (devitrified means a glass in which small
crystals have formed and are held in the glassy structure).

Many crystalline solids can be produced (and indeed used) as rather large
crystals with the order preserved across the whole crystal (up to several
centimetres across). These are called 'single crystals'. Generally, however,
solids consist of agglomerates of small crystallites or *grains* (usually 0.01
to 1.0 mm across). Such a solid is described as 'polycrystalline'.

crystal grains, or crystallites

The change to the solid state usually starts at many points in the freezing
liquid. Growth proceeds at each point until the individual crystals (grains)

touch and all of the liquid has solidified. Very careful control of the solidification process is required to produce a single crystal. This is achieved in the manufacture of transistors†, lasers† and quartz oscillators†.

As is shown schematically in Figure 10, the grains in a polycrystal are differently aligned with respect to one another. Notice, for example, that the hexagons drawn on the two halves of the Figure are tilted with respect to each other. In the regions where neighbouring grains are in contact (*a grain boundary*) the atoms are in intermediate positions between the regular atomic arrays in the two grains. Polycrystalline structure on a large scale can be seen, for example, on galvanized dustbin lids and on 'iced-up' windows. A grain boundary is another type of crystal imperfection.

The presence of various types of crystal imperfection has many very important consequences in the structure and, therefore, the properties of materials. Details would not be appropriate at this stage, but a few general examples will be considered.

Although the number of vacancies present in a crystal is very small (about one in 10 000 atoms at the melting point), they are the means by which atoms can move around in crystals, and atom migration of this kind is important in things as diverse as, for example, the oxidation† of metals and the control of the electrical properties of transistors.

The prime reason why metals are so useful is that they can be made to undergo a permanent change of shape, *plastic deformation*, by means of a mechanical load (a pull, for example). Consequently they can be fabricated into complex shapes, though when in service as a constructional material this property can also be a nuisance, because it may occur with time as metal fatigue. Plastic deformation usually involves the sliding of crystal planes over one another, rather like that which occurs when a pack of cards is 'sheared' (Fig. 11). This produces the change of shape.

The sliding process takes place much more easily in metals than in, say, diamond. Can you suggest any reasons why this is so?

Different types of bonding hold the atoms in place in the crystals of these substances (Unit 8). In diamond, the bonding is of a 'directed' nature, because it is in a specific direction between two specific carbon atoms. By comparison, the bonding in metals is 'non-directional', occurring indiscriminately between many atoms which therefore are able to slide over one another relatively easily.

When it is easier for cracks to open and grow than for plastic deformation to occur, a solid breaks. In contrast with metals, glasses and ceramics (such as brick and china) crack more easily than they deform plastically and are, therefore, brittle. Polymers are usually much softer than metals, because the bonding forces are weak, so that the molecules slide over one another very easily.

Grain boundaries have a marked effect on plastic deformation and fracture because they influence the sliding which can occur in the individual grains and the ease with which the cracks spread. Grain boundaries and other crystal imperfections can influence strongly the magnetization process in magnetic materials (Unit 23).

If, after reading this section, you wish to learn more about crystal structure, you can refer to the section on 'The internal structure of minerals' in Chapter 1 of the Earth Sciences Reader, *Understanding the Earth.*

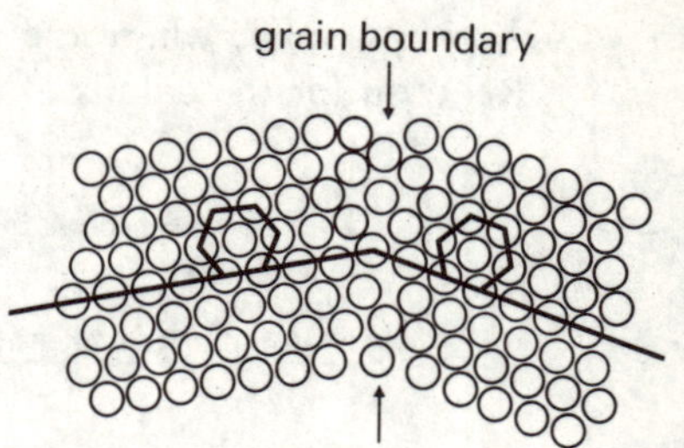

Figure 10 Portions of two grains in a polycrystal differently aligned, as is apparent from the two hexagons drawn on the regularly packed atoms. In the grain boundary between the grains atoms are in intermediate positions between the two regular atomic arrays.

plastic deformation

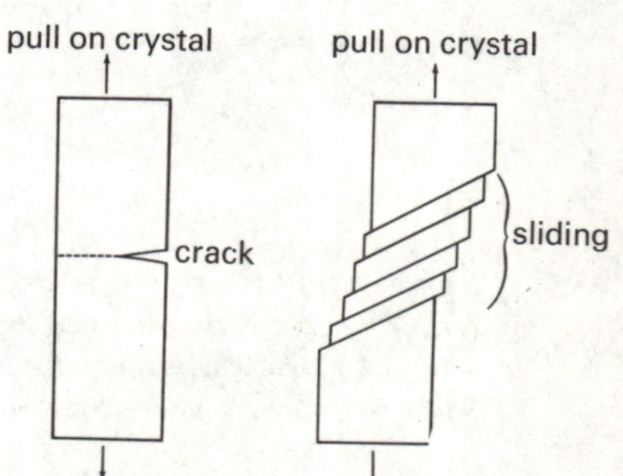

Figure 11 Two responses to a tensile force (a pull) on a crystal. The crystal may crack, or it may elongate by a sliding process.

Question 6 (*Objective 6*)

Two gases have the same temperature. This means that the (select from a—f) of the molecules in the two gases are equal, and this is a manifestation of the (select from g—i) law of thermodynamics.

(a) velocities   (b) average velocities   (c) masses

(d) momenta   (e) kinetic energies   (f) average kinetic energies

(g) Zeroth   (h) First   (i) Second

Question 7 (*Objectives 2 and 5*)

Simple kinetic theory assumes that (select as many as appropriate from a—f)

(a) The molecules are negligibly small.

(b) The molecules attract each other.

(c) Pressure is the energy of the molecular collisions.

(d) All collisions are perfectly elastic.

(e) All gases are ideal.

(f) The molecules slow down in the course of time.

Question 8 (*Objectives 2 and 12*)

(i) 0.016 kg of oxygen occupy 22.4×10^{-3} m³ at standard atmospheric pressure (10^5 N m^{-2}) and zero degrees Centigrade (273 K). What is the average $\overline{v^2}$ (squared velocity) of the molecules in m sec^{-1}? Select from a—g.

    (a) $\dfrac{3 \times 16 \times 10^5}{22.4}$

    (b) $\dfrac{3 \times 16 \times 10^3}{22.4}$

    (c) $\dfrac{3 \times 22.4 \times 10^5}{16}$

    (d) $\dfrac{3 \times 22.4 \times 10^3}{16}$

    (e) $\dfrac{16 \times 22.4 \times 10^5}{3}$

    (f) $3 \times 16 \times 22.4 \times 10^5$

    (g) $3 \times 16 \times 22.4 \times 10^3$

(ii) If the temperature is lowered to 200° K, the average squared velocity will (h) increase, (i) decrease, (j) stay the same? (Select from h—j).

Question 9 (*Objective 13*)

It is easier to unscrew a metal top from a glass bottle after running hot water over them, because (select one from a—e).

(a) The heat destroys the vacuum.

(b) Glass contracts under these conditions.

(c) The pressure inside the bottle increases.

(d) Metal expands more than glass when heated.

(e) The contents of the bottle expand.

Question 10 (*Objective 9*)

Make lists of the bulk properties of a solid, a liquid and a gas, both those properties mentioned in this Unit and any extra properties from your general knowledge. Compare with the list made earlier (p. 7).

Question 11 (*Objective 9*)

Fill in the matrix below with the words Yes/No (columns 1, 3, 4) or Important/Unimportant (columns 2 and 5).

	1 Do the molecules have a fixed position?	2 Are the interactions with other molecules important or unimportant for the bulk properties?	3 Does the material have a definite shape?	4 Does the material have a definite volume?	5 Are the interactions of the molecules with the walls or boundaries important or unimportant for the bulk properties?
(An ideal) GAS					
LIQUID					
SOLID					

Question 12 (*Objective 7*)

State Avogadro's Law in words, making certain that you are clear about the conditions in which it holds (for example, what temperature conditions hold). Then using these conditions, and the ideal gas law ($PV = nkT$) show that Avogadro's Law is a necessary consequence.

Question 13 (*Objective 5*)

Derive Boyle's Law, in the form $PV = \frac{1}{3}nm\overline{v^2}$, using simple Newtonian mechanics.

Question 14 (*Objective 10*)

Each of a—f below is a verbal or mathematical statement. Convert the verbal ones to mathematical ones and vice versa.

For example:

1. *Given*: the area, A, of a circle is given by squaring the value of the radius, r, and multiplying it by a constant, π.

Answer: $A = \pi r^2$.

2. *Given*: $l_2 = l_1 (1 + \Delta T.c_e)$ where l_1 is the length of a metal rod, c_e its coefficient of linear expansion, and ΔT is a small increase in the temperature of the rod.

Possible Answer: If a rod, of length l_1, is heated through ΔT degrees, its length is increased to l_2 by an amount equal to l_1 multiplied by the product of the number of degrees and its coefficient of linear expansion.

(a) The density, D, of a body is defined as its mass, M, divided by its volume, V.

(b) The pressure, P, in a fluid is defined as the force, F, per unit area of an area, A.

(c) A rectangular block of metal (dimensions x, y, and z and density D) weighs M_w in water and M_a in air. The difference is equal to the weight of water displaced by the metal. The density of water is D_w.

(d) $P_1 V_1 = P_2 V_2$ where P and V are pressure and volume of a gas respectively.

(e) $T \propto (\overline{KE})$ for an ideal gas.

(f) The ideal gas law states that for a given amount of an ideal gas, the product of pressure and volume is proportional to the temperature.

Question 15 (*Objective 2*)

Re-read the objective, and then arrange the expressions and 'connectives' listed below to form a simple coherent description apposite to the objective.

You can use any term more than once.

Refer to section 5.3.2 of the Unit if you have doubts about the adequacy of your answer or the level of complexity we refer to as a 'simple description'.

Expressions		*'Connectives'*	
1	force per unit area	12	called
2	elastic collisions	13	and
3	exchange of momentum	14	with
4	walls of the container	15	the
5	molecules of a gas	16	leads to
6	collision with the walls	17	make
7	rate of change of momentum	18	at
8	holding the gas	19	is
9	other molecules		
10	pressure of the gas		
11	represents a force		

Question 16 (*Objective 1*)

  (i) An *elastic collision* is
      (a) when two rubber balls collide,
      (b) a head-on collision with a wall,
      (c) a collision where kinetic energy is conserved.

  (ii) *Temperature* is
      (a) the sensation of hotness,
      (b) a measure of the kinetic energy of the molecules,
      (c) the force of molecular motion.

 (iii) *Avogadro's number is*
      (a) the number of molecules in one mole,
      (b) the number of molecules in the world,
      (c) the number of molecules in 1 m^3 of hydrogen at absolute zero.

  (iv) *Cohesion* is
      (a) the tendency of molecules to stick together,
      (b) the change of state from liquid to solid,
      (c) the force which acts in the surface of a liquid.

Question 17 (*Objective 1*)

Define in your own words:

  (i) intermolecular forces,
 (ii) a crystalline array,
(iii) a glass.

5.6 Changes of State

You are familiar, from your everyday experience, with the different states of matter. There are gases (the air), liquids (your bath water) and solids (the floor). You have listed some of their properties in your answer to *SAQ* 10 above, and you now know some molecular differences between the states. But you are also familiar with instances of *change of state*: water *boils* to steam, steam *condenses* to water, ice *melts* to water. If a wet garment is hung out on a cold day, the water on it may first *freeze*, and then *sublime* directly to water vapour without melting. All these—melting, freezing, subliming, boiling—are changes of state. In this section you will look at these changes from a molecular point of view.

> **Can you think of one concept that sums up the difference between the three states of matter?**
>
> **(Look at the answer to *SAQ* 11 (p. 46), and try to infer this.)**

In this Unit the states of matter have been examined by implication, though not until now explicitly, in terms of *increasing amounts of order*, where the term 'order' means the degree to which the relative positions of the molecules can be specified during the course of time. The molecules in a gas have no definite position, their movement is limited only by the walls of the container in which they are placed. Their interactions with each other give rise to rather small net effects; their interactions with the walls of the container are very important for the total properties of the gas. The molecules in a liquid are always in contact, because they are bound to each other by cohesive (attractive) forces, but they have no fixed position within the definite volume of the liquid. This is a different state of order from that in the gaseous state. Again, the molecules in a solid are bound at fixed positions in the volume of the solid. They may vibrate about these positions, but their average position within the solid stays the same. Clearly then, the changes of state between gas and liquid and solid are points at which the degree of ordering of the molecules changes. Under a definite pressure, the points at which liquid changes to gas or solid changes to liquid have particular temperatures. These are called respectively, the boiling point and the melting point. The order changes at these points and energy must be supplied to the system to overcome the binding forces holding the molecules in the lattice of the solid or the cohesive forces on the liquid molecules. There is a definite amount of energy required to change a given mass of solid into liquid or liquid into gas. This amount of energy is called, in the case of melting, the *latent heat of fusion** of the material, and in the case of boiling, the *latent heat of vaporization*.*

order

latent heats of fusion and vaporization

In both these cases, the degree of order of the molecule is changing; in fact a function can be defined, the *entropy* of the system, which can be considered as a measure of the order (or rather of the disorder, since systems with greater order have a smaller entropy). See, however, the qualification made at the beginning of section 5.6.1.

entropy

> **Do you understand the use of the term 'function' in the last sentence?**

Function in the sense used here means some property which can be defined in a precise way, and which has a value which can be measured,

---

* *Fusion and vaporization are alternative terms for melting and boiling respectively.*

at least in principle. Thus, for a given mass of gas, the pressure, the volume, the temperature, are all functions measuring precise properties (the temperature, for example, measures the average kinetic energy of the gas molecules).

5.6.1 Entropy and the Third Law of Thermodynamics

It is not easy to give a precise definition of entropy. A correct definition involves mathematics outside the scope of this course. The basis of the definition is the probability of finding a system in a particular state: the more improbable the state, the lower the entropy. In the simple systems that are considered in this Unit, probability can be loosely equated with disorder, but this is an analogy which must not be pressed too closely with all systems. It does, however, serve to give a first feeling for the entropy concept.

In the case of changes in state, a simple measure of the change of entropy is available, given by the equation:

$$\text{Change in entropy } (\Delta S) = \frac{\text{Latent Heat } (L)}{\text{Temperature } (T)}$$

This equation, which is here arbitrarily stated without proof, describes the change of entropy under certain specified circumstances. The entropy of the gas is *greater* than that of the liquid, the entropy of the liquid is *greater* than that of the solid.

Do you expect the entropy changes on melting or in vaporization to depend on the material which is melting or vaporizing, or would you expect them to be the same for all materials?

The answer to this question is not simple. Different materials are likely to be very different in their physical properties. However, the change in order from, say, liquid to gas might be very similar for all liquids, so that the entropy change might also be similar. In fact, many liquids, with very different latent heats and boiling points, have a ΔS of about 90–100 J mol^{-1}K^{-1} when changing state from liquid to gas. Exceptions include water and alcohol. This generalization, which is called Trouton's Rule, is exemplified in Table 1.

Table 1

Substance	Latent Heat (L) (J mol^{-1})	Boiling point (K)	L/T (J mol^{-1}K^{-1})
Ethyl ether	27 200	307	89
Benzene	31 500	353	89
Propylacetate	34 800	373	93
Mercury	59 600	630	95
Zinc	115 000	1 180	98
Water	40 700	373	109
Ethyl alcohol	39 700	351	113

Notice that in this table, to compare the same number of molecules of each substance, the latent heats are given per mole *rather than* per kilogramme.

The equation $\Delta S = L/T$ gives the change in entropy in a particular set of circumstances, that of change of state. It is possible to calculate entropy changes under many other specified circumstances (for example, on mixing two different gases), but this will not be attempted here.

In order to put an absolute value on the function entropy (note that so far only *changes of entropy* have been discussed), it is necessary to have some definite zero point of entropy. (Just as a zero point of temperature was necessary in order to put an absolute value on temperature.) Now it happens that the absolute zero of temperature is also the zero point of entropy. This is the Third Law of Thermodynamics. You may think at first sight that it is the energy of the molecules which should become zero at absolute zero, but quantum mechanics (Units 29 and 30) shows that this is not the case; rather it is the order of the molecules which becomes perfect and, therefore, the entropy becomes zero. At, or near, absolute zero the molecules may still have a quantum-mechanical zero-point energy, giving them many fascinating properties at very low temperatures, such as superconductivity† and superfluidity†, which those of you who continue in the physics field will hear about in later years.

Notice that the zero of entropy is a point of perfect order and that in general the greater the entropy of a system the more is the amount of disorder in that system. The Demon in Maxwell's thought experiment was trying to decrease the entropy of the gas molecules, that is to increase their order, without expending energy, and this he was unable to do because of the Second Law of Thermodynamics. (If you do not remember the Second Law of Thermodynamics, refer back to section 5.3.4.)

Any process which enables us to specify more closely where a certain molecule can be found will clearly increase the order; to say that in a given mass of gas on one side of a certain plane all the slow molecules will be found and on the other side of the same plane all the fast molecules will be found is an increase in specificity and, therefore, an increase in order compared with the general case, when all the fast and slow molecules are equally distributed throughout the whole mass. To separate the molecules in this way is to decrease their entropy.

Figure 12 represents a container with a removable partition, which divides it into two equal parts A and B. Initially A contains a certain number of atoms of the inert gas neon and B the same number of atoms of the inert gas helium. The partition is removed and the gases mix. No chemical reaction takes place and the temperature does not change. Does the entropy in this process increase or decrease, or does it stay the same?

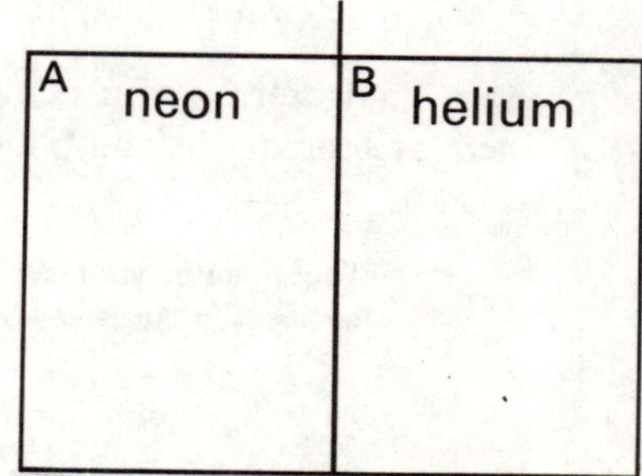

Figure 12

It increases. The state before the experiment, with all the neon atoms on side A and all the helium on side B is clearly more closely specified, and therefore more ordered, than the state after when the two sorts of atoms are mixed through the total volume. An increase in entropy goes with an increase in disorder. If you had difficulty in answering this question you may not have understood the last section completely. Read on, and it should become more clear. For simplicity, suppose there are 10 atoms on each side of the wall, and suppose all the neon atoms are marked with an A and all the helium atoms with a B. Work out the number of different arrangements of A and B atoms at any time *after* the partition is removed. Assume there are still 10 per side, for Avogadro's Law shows that there is no reason why the number of particles per unit volume should be different on the two sides. The possible arrangements are:

Side A	Side B
10A	10B
9A 1B	1A 9B
8A 2B	2A 8B
7A 3B	3A 7B
6A 4B	4A 6B
5A 5B	5A 5B
4A 6B	6A 4B
3A 7B	7A 3B
2A 8B	8A 2B
1A 9B	9A 1B
10B	10A

So there are *eleven* possible arrangements and in only two of them (the first and last as written here) are the two sorts of molecules separated on the two sides of the original wall. Think how many more possible arrangements there would have been if there were 10^{23} atoms on each side originally. Again only in two would the atoms be separated.

Clearly, then, the arrangement where the atoms of the two types are separated is a very improbable state representing a *higher degree of order* than states where there is a mixture on both sides. Conversely, the mixture is a more probable state, or a *lower degree of order*, and thus has a *higher entropy*, than the separated state.

Note that in the argument above it was assumed that all the atoms A were equivalent. In fact each atom should be considered separately. This does not affect the first and last arrangement, where all 10A and 10B are on opposite sides, but in the case, for example, where there are 9B and 1A there are ten different ways of choosing the one A atom. There are even more ways of choosing the 2A atoms in the arrangement of 8B and 2A. Because of this the *total* number of different ways of dividing up the atoms is very large (in fact, $\sim$185 000) and still only two of these have atoms A and B separated on opposite sides.

As a final exercise you could take a fresh box of matches and upturn them at shoulder height, allowing the matches to fall on the floor.

Would you be surprised if the fallen matches are arranged in ranks with all the heads in the same direction, like parading soldiers?

Yes, you would be very surprised!

Would this process increase, or decrease, the entropy of the matches?

Surely you answered this one correctly!

We shall be returning to this topic in the television programme that is to be presented in the 'Revision Week' following Unit 13. Among other things, we shall be showing how these ideas about probability and disorder are related to the idea of *direction in time*—to 'the arrow of time'.

5.7 Water—an Eccentric Liquid*

You have seen that liquids in general increase in volume as they are
heated. Water at temperatures less than 4° C is, however, an exception to
this rule. Substances generally have a smaller volume per unit mass in the
solid state than in the liquid state at the melting point, but it is well known
that water expands on freezing and causes pipes to burst. Biologically it
is of great importance that water expands in this way; it means that ice
floats on top of water and seals the bulk of the water from further freezing.
Ice conducts heat poorly and forms an insulating layer, like lagging on a
tank. If ice were more dense than water and sank to the bottom, a continual
process of freezing would take place which might turn certain areas of the
world into ice wastes. When ice melts, it contracts by about 10 per cent,
and as the temperature is increased the contraction increases for a further
4° of temperature rise.

Table 2

Density of water at various temperatures

Temperature ° C	State	Density kg m^{-3}
0	Solid	917.0
0	Liquid	999.8
3.98	Liquid	1000.0
10	Liquid	999.7
25	Liquid	997.1
100	Liquid	958.4

At 4° C, the water has a maximum density or minimum volume; thereafter
it expands like any other liquid as the temperature is increased. Water
also departs markedly from Trouton's Rule (see section 5.6.1 above).
The value of ΔS from water to steam is 109 J mol^{-1}K^{-1}, suggesting that
the change in order is greater than in the evaporation of a normal liquid
and thus that water has extra order. (Trouton's Rule value for normal
liquids is 90–100 J mol^{-1}K^{-1}.)

Röntgen, the discoverer of X-rays, was fascinated by these strange
properties of water. Many years ago he predicted correctly that the
structure of ice was more open than that of water above the freezing
point; that is the molecules were further apart in the solid (crystalline)
structure than in the liquid. He further suggested that some of the open
structure of ice survived in the liquid water above the freezing point, so
that water just above freezing point was effectively a mixture of two types
of liquid, an 'ordered' liquid having some of the open structure of ice and
another liquid having a more normal, more disordered and more close
structure. Röntgen supposed that the change in the properties of water as
the temperature was increased depended on the different amounts of
these two sorts of water that were present at any particular temperature;
that as the temperature increased the proportion of ordered water would
decrease and the proportion of normal water would increase. It is now
known that the special nature of water, the expanded structure of ice and

* *See also Unit 8.*

the persistence of ordered regions in water above the freezing point, all depend on the type of chemical bonding which occurs between water molecules (see Figs. 13a–c which, however, are purely diagrammatic and very much simplified).

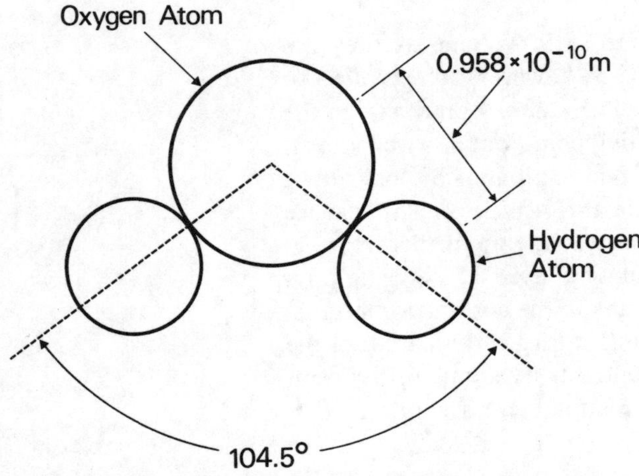

Figure 13 (*a*)   *An idealized diagram of the water molecule, one oxygen atom bonded to two hydrogen atoms.*

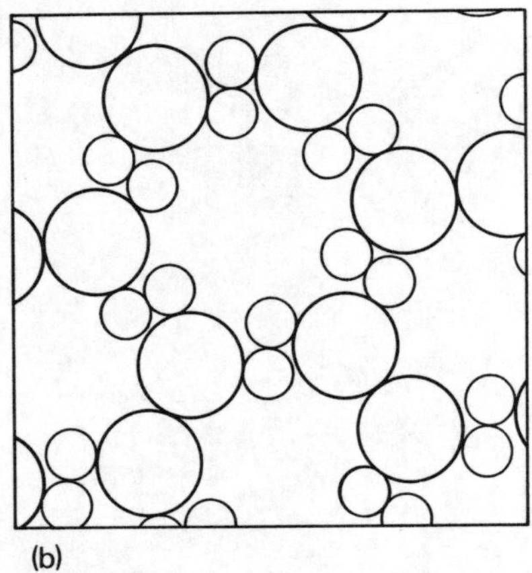

(b)

Figure 13 (*b*)   *In ice the water molecules (see Fig.* 13 (*a*)) *have regular lattice positions. (This is a highly idealized diagram, and should not be taken to represent real positions of hydrogen and oxygen atoms. The point is simply that the structure is regular, and that the atoms stand further apart than in the liquid.)*

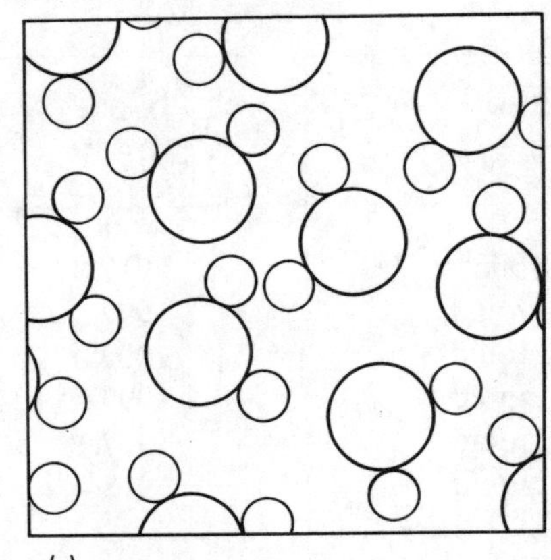

(c)

Figure 13 (*c*)   *In liquid water there is much less regularity of structure. As you have read in the text some regions of 'ordered' water-structure persist in the liquid. (Again a highly idealized diagram.)*

A more detailed picture of the normal ice lattice is given in Figure 14, where the three-dimensional structure is apparent. Notice that each oxygen atom is surrounded by four other oxygen atoms, and that on each line joining oxygen to oxygen there is a hydrogen atom close to its parent oxygen atom but attracted electrostatically by the oxygen of a neighbouring atom. In the water molecule, the hydrogen atoms have a small positive charge and the oxygen atoms have a small negative charge, making this a weak electrostatic bond (called a hydrogen bond, but the name is not important at this stage). Because this electrostatic bond forms when the three atoms are in line or nearly in line, the ice structure, which contains a maximum number of these bonds, is more open than the disordered water structure, where the positions of the hydrogen atoms are not subject to this constraint. In water, the individual molecules can fit more closely together, because the thermal energy is too high for so many of the electrostatic bonds to persist. The details of chemical bonding are a matter for later study (Units 8 and 9), but it may be remarked at this

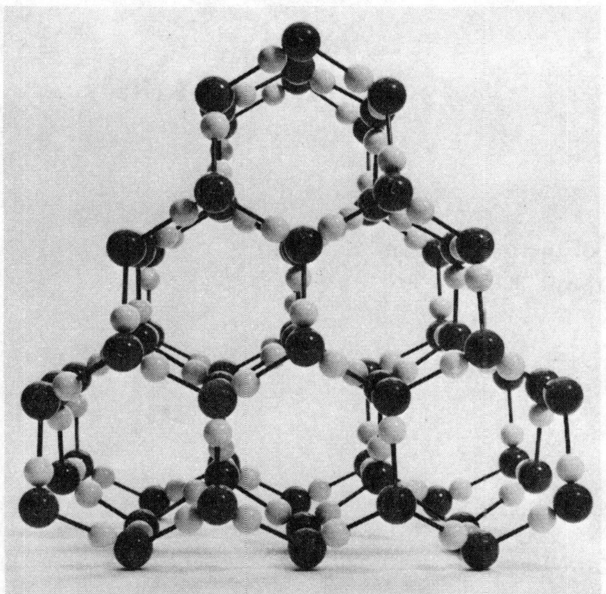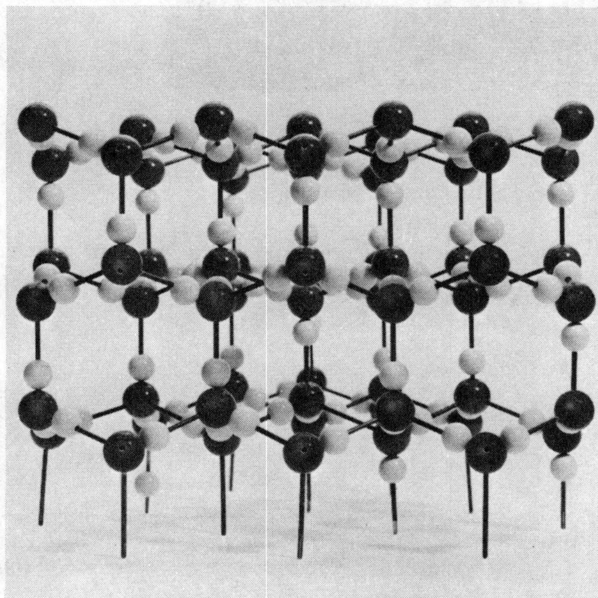

Figure 14   Two views of the normal ice lattice. The black balls represent oxygen atoms and the white balls represent hydrogen atoms. Notice that each oxygen atom is surrounded symmetrically by four other oxygen atoms forming the points of a regular tetrahedron. Along each line joining the oxygen atoms there is a hydrogen atom which is closer to one oxygen atom than to the other. The hydrogen atom is closer to the oxygen atom which forms part of the same water molecule, but is bonded electrostatically to the other, more distant, oxygen atom. In this way each oxygen atom has two hydrogen atoms close to it forming part of the same water molecule, and two more hydrogen atoms further from it but bonded to it electrostatically.

point that the consequences of the special nature of water are widespread. For example, ordered regions in water persist at normal room temperatures and are known to surround large molecules, the molecules in living cells for example. Though the extent of these ordered regions is a matter of current debate, it seems very likely that the special nature of water is of vital concern in the modern biological field and in the very nature of life itself. These points will be discussed in this Unit's radio programme, and you are asked to have Figure 14 at hand when you listen to it.

Book List

Other accounts of some aspects of heat, thermodynamics and kinetic theory are given in:

(a) V. L. Parsegian, *et al., Introduction to Natural Science, Part 1: The Physical Sciences.* Academic Press, 1968. Chapter 10.
(b) E. R. Huggins, *Physics 1.* Benjamin, 1968. Chapter 10.
(c) K. Krauskopf and A. Beiser, *Fundamentals of Physical Science,* 5th edition, McGraw-Hill, 1966. Chapter 9.
(d) J. Orear, *Fundamental Physics,* 2nd edition. John Wiley, 1967. Chapters 5 and 6.

Densities and Pressures

The *density* of a material is the mass of unit volume of the material. In the SI system densities are expressed as kg m^{-3}. Typical densities are given in Table 4.

density

Table 4

kg m^{-3}

Water	1×10^3 (*but see also Table 3*)
Mercury (liquid)	13.6×10^3
Wood (oak)	0.6–0.9×10^3
Steel	7.8×10^3
Lead	11.3×10^3
Sea-water	1.025×10^3
Muscle	$\sim 1.1 \times 10^3$
Fat tissue	$\sim 0.9 \times 10^3$
Bone	~ 1.5–1.7×10^3

Pressure is defined as the force acting on unit area, and is measured as newton m^{-2}. In any fluid (liquid or gas) the pressure is, at any point, the same in all directions. In a liquid, the pressure of the liquid is given, at any distance d meters below the surface, by $\rho.d.g$, where ρ is the density of the liquid (see above) and g is the acceleration due to gravity (see Units 3 and 4).

pressure

To this must be added the pressure of the atmosphere itself, which is normally acting on the surface of the liquid. Standard, or normal, atmospheric pressure will support a column of 0.76 m of mercury (or 10.3 m of water) in a barometer tube. (If you do not know how a mercury barometer works look it up in any simple physics textbook.) By the formula already given (remember $g = 9.81$ m sec^{-2})

Normal atmospheric pressure $= 0.76 \times 13.6 \times 10^3 \times 9.81$ newtons m^{-2}

$$\approx 10^5 \text{ newton m}^{-2}.$$

(In the old British system of units, normal atmospheric pressure is about 14 pounds (weight) per square inch, or p.s.i.)

When a body is immersed in water (or any liquid), it will displace some of the liquid. The apparent weight of the body will be the weight of the body in air less the weight of the liquid displaced. Suppose the body is volume V, and density ρ_s, and the liquid is density ρ_l.

Then the weight (the force of gravitation acting on the body, remember) in air is $v . \rho_s . g$ and the weight in water, assuming that the body sinks, is $v(\rho_s - \rho_l) g$.

This is *Archimedes' principle*, which may be stated, 'The upthrust (this means the difference between the weight in air and in liquid) on a body in a liquid is equal to the weight of liquid displaced.'

Archimedes' principle

If the density of the solid is less than that of the liquid it will float.

This will also happen if the solid is in the form of a thin, liquid-tight, skin and is filled with lighter gas, the air for example. When a body such as a ship floats in water, the weight displaced by the ship must, by Archimedes' principle, be equal to the weight of the ship itself (measured in air, which would require a rather large pair of scales for direct measurement). You will sometimes hear reference to a ship having a (water) 'displacement' of so many tons (weight).

All these points are covered in detail in almost any elementary science textbook which you can find in your public library.

Appendix 2

Glossary

LASER   A device for producing a high-intensity beam of monochromatic, coherent light.

INERT GAS   A gas which does not take part in any chemical reaction.

OXIDATION   A chemical reaction with oxygen—often the oxygen in the air. Rusting is an example of oxidation.

QUARTZ OSCILLATOR   An oscillator whose frequency is controlled by the mechanical and electrical oscillation of a carefully-ground plate of crystalline quartz.

SUPERCONDUCTIVITY   At very low temperatures (1–2K) some metals (e.g. lead) become able to conduct electricity with almost no loss of electrical power. This property is sometimes used in high-power electro-magnets.

SUPERFLUIDITY   At very low temperatures (1–2K) liquid helium (an inert gas) flows easily through obstructions, and against gravity.

TRANSISTOR   A device using crystals of germanium or silicon, with impurities which change the electrical properties. A junction between two such crystals can be made to amplify (increase) electric currents. Modern electronics is greatly dependent on transistors of all types.

Self-Assessment Answers and Comments

Question 1

The correct answer is (c).

This is an application of $PV = RT$

Since V and R are constant we can write

$$\frac{P_1}{P_1} = \frac{T_2}{T_2}$$

Remember that T_1 and T_2 must be in degrees Kelvin.

So
$$P_2 = 250 \times \frac{(273 + 37)}{(273 + 17)}$$

Question 2

The correct answers are (i) c and (ii) g.

$17°$ C is 17 Centigrade degrees above melting point.

This is $17 \times \dfrac{180}{100}$ Fahrenheit degrees above melting point. But melting point

is already $32°$ on the Fahrenheit scale. So the answer is $32 + (17 \times 180/100)$.

Similarly for $37°$ C, the Fahrenheit equivalent is $32 + (37 \times 180/100)$.

Question 3

72 m of water is approximately (7×10.3) m, and you will remember from Appendix 1 that 10.3 m of water* is equivalent to a pressure of one atmosphere, so that pressure *due to the water* is about 7 atmospheres. But the surface of the water is already at a pressure of one atmosphere, so that total pressure at the wreck is 8 atmospheres.

Use
$$PV = RT, \text{ here } R \text{ and } T \text{ are constant}$$

so
$$P_1 V_1 = P_2 V_2$$

$$V_2 = \frac{P_1 V_1}{P_2} = \frac{8 V_1}{1}$$

If the *volume* of the bubble increases by a factor of eight the *diameter* must increase by the *cube root* of eight, or 2.

Question 4

The correct answer is (d).

The air will pass into the tyre when the pressure in the pump is just greater than that in the tyre. If the piston is one third down the barrel the air in the pump is compressed by a factor of

$$\frac{1}{\frac{2}{3}} \text{ or } \tfrac{3}{2}$$

The pressure in the pump will be about $\tfrac{3}{2}$ atmospheres, or 21 p.s.i.

Strictly speaking, for sea water, more dense than fresh water, this is $(10.3/1.025)$ m. For an approximate answer you can neglect this factor.

Question 5

The correct answer is (e).

Here the length of the gas column is proportional to the volume of the gas, since the diameter is uniform.

Use $\qquad PV = RT \qquad R$ and T are constant

so $\qquad \dfrac{V_1}{T_1} = \dfrac{V_2}{T_2}$

giving $\qquad \dfrac{12.4}{(273)} = \dfrac{l}{(273-39)} \quad$ where l is the required length

(Remember, Kelvin temperatures are necessary.)

Question 6

The correct answers are (i) f and (ii) g.

If you did not get (i) correct, read section 5.3.2 again, particularly page 19.
If you did not get (ii) correct, read section 5.2.2 again.

Question 7

The correct answers are (a) and (d).

(b) is true, but is disregarded in *simple* kinetic theory, where it is assumed that there are no molecular attractions.

(e) and (f) are not true, nor are they assumptions of the theory.

(c) is nonsense.

If you had difficulties, re-read sections 5.3.2 and 5.3.5.

Question 8

The correct answers are (i) c and (ii) i.

(i) Use $\qquad PV = \tfrac{1}{3} nm\overline{v^2}$

so $\qquad \overline{v^2} = \dfrac{3\,PV}{(nm)}$

(nm) is the total number of molecules in volume V multiplied by the mass of one molecule, so is the total mass of gas present, 0.016 kg as given. P and V are given, again in SI units.

So $\qquad \overline{v^2} = \dfrac{3 \times 10^5 \times 22.4 \times 10^{-3}}{0.016}$

which simplifies to (c).

(ii) You should have had no difficulty with this, re-read 5.3.2 if necessary. In fact $\overline{v^2}$ will decrease in the ratio of the two Kelvin temperatures. (Remember $T \propto (\overline{\text{KE}})$.)

So the new $\overline{v^2}$ would be $(c) \times \dfrac{200}{273}$

Question 9

The correct answer is (d).

(a) and (b) are not true.

(c) and (e) are true, but not relevant.

Question 10

Some bulk properties of a solid, a liquid and a gas.

A Solid	A Liquid	A Gas
Definite shape and form	Definite volume, but no definite shape	Fills all available volume
Can withstand stresses, compression tension, twist, etc.	Can withstand compression, but cannot withstand tension	Can easily be compressed
Malleability brittleness resilience, i.e. 'structural properties'		
Does not flow (though it may settle slowly under gravity)	Flows, the ease of flow is measured by the 'viscosity'	Flows easily (i.e. low viscosity)
	Can dissolve solids and gases, which then diffuse more or less easily through the liquid	

These are just some of the more important properties—the question is open ended and you can write in further properties as you proceed in your course.

Question 11

The correct matrix is given below.

The brackets in the last two items in column 5 were inserted because there are properties, which you have not yet met, which are concerned with these interactions. Surface tension of a liquid is an example.

	1 Do the molecules have a fixed position?	2 Are the interactions with other molecules important or unimportant for the bulk properties?	3 Does the material have a definite shape?	4 Does the material have a definite volume?	5 Are the interactions of the molecules with the walls or boundaries important or unimportant for the bulk properties?
(An ideal) GAS	No	unimportant	No	No	important
LIQUID	No	important	No	Yes	(unimportant)
SOLID	Yes	important	Yes	Yes	(unimportant)

This question is central to the whole Unit. If you got the wrong answers or had difficulty in answering, you should re-read sections 5.3–5.6 inclusive, and consult some of the textbooks in the reading list if necessary.

Question 12

The answers to both parts of this question are given in section 5.3.3. Refer there if necessary.

Question 13

This derivation is given in section 5.3.2, pages 17–20.

Question 14

(a) $D = \dfrac{M}{V}$

(b) $P = \dfrac{F}{A}$

(c) $(M_w - M_A) = x \cdot y \cdot z \cdot D_w$

(d) The product of the pressure and the volume of an (ideal) gas under the first condition is equal to the same product for the same mass of gas under the second condition. (This is true under constant temperature conditions.)

(e) The temperature of the ideal gas is proportional to the average kinetic energy of the molecules.

(f) $PV = RT$

Question 15

Suggested answer:

Molecules of a gas make elastic collisions with other molecules and with
the walls of the container holding the gas. Collision with the walls leads
to exchange of momentum, and the rate of change of momentum at the
walls of the container represents a force on the walls of the container.
The force per unit area on the walls of the container is called the pressure
of the gas.

Question 16

  (i) The correct answer is (c).
     Defined in section 5.3.1, page 15.

 (ii) The correct answer is (b).
     Explained in section 5.3.2, page 19.

(iii) The correct answer is (a).
     Explained in section 5.3.3, page 20.

(iv) The correct answer is (a).
     See sections 5.3.5 and 5.4, pages 22–24.

Question 17

  (i) Intermolecular forces are forces between molecules. They may be
     attractive (cohesive) or they may be forces of repulsion (for example,
     the force between two molecules with electric charges of the same
     sign).

 (ii) A regular array of atoms or molecules in a crystal. This array has
     long range order (see page 25).

(iii) A solid in which there is no regular crystalline array, but where the
     molecular positions are randomly arranged, without long range
     order (see page 27).

Acknowledgements

Grateful acknowledgement is made to the following sources for
illustrations used in this Unit: D. C. HEATH AND COMPANY, Lexington,
Mass., U.S.A., from *PSSC Physics*, 1965, Fig. 3; DR. PAUL BARNES and
MR. IAN CHERRY, Department of Crystallography, Birkbeck College,
London, Fig. 14.